Other Publications in the CECD Series

S&T Revitalization: A New Look
Davinder K. Anand, Lisa M. Frehill, Dylan A. Hazelwood, Robert A. Kavetsky, Elaine Ryan
ISBN 978-0-9846274-5-5

Rare Earth Materials. Insights and Concerns
Michael G. Pecht, Robert E. Kaczmarek, Xin Song, Dylan A. Hazelwood, Robert A. Kavetsky, and Davinder K. Anand
ISBN 978-0-9846274-4-8

Energetics Science and Technology in Central Europe
Invited Editor: Ronald W. Armstrong
Series Editors: James M. Short, Robert A. Kavetsky, and Davinder K. Anand
ISBN 978-0-9846274-3-1

Simulation-Based Innovation and Discovery. Energetics Applications
Edited by Davinder K. Anand, Satyandra K. Gupta, and Robert A. Kavetsky
ISBN 978-0-9846274-2-4

Other Publications in the CECD Series

Energetics Science and Technology in China
James M. Short, Robert A. Kavetsky, Michael G. Pecht, and Davinder K. Anand
ISBN 978-0-9846274-0-0

From Science to Seapower.
A Roadmap for S&T Revitalization.
Postscript 2010
Robert A. Kavetsky, Michael L. Marshall, and Davinder K. Anand
ISBN 978-0-9846274-1-7

Training in Virtual Environments.
A Safe, Cost-Effective, and Engaging Approach to Training
Satyandra K. Gupta, Davinder K. Anand, John E. Brough, Maxim Schwartz, and Robert A. Kavetsky
ISBN 978-0-9777295-2-4

From Science to Seapower.
A Roadmap for S&T Revitalization.
Robert A. Kavetsky, Michael L. Marshall, and Davinder K. Anand
ISBN 978-0-9846274-1-7

Engineering For Social Change

Engineering Is Not Just Engineering

Center for Engineering Concepts Development Series

Engineering For Social Change

Engineering Is Not Just Engineering

Davinder K. Anand
Dylan A. Hazelwood
Michael G. Pecht
Mukes Kapilashrami

Center for Engineering Concepts Development Series

University of Maryland, College Park, Maryland

Cover concept and design by Davinder K. Anand and Dylan A. Hazelwood

Library of Congress Control Number: 2016945417

Direct all inquiries to:

Center for Engineering Concepts Development (CECD)
Department of Mechanical Engineering
2140 Glenn L. Martin Hall
University of Maryland
College Park, MD 20742
Ph.: (301) 405-5205
http://www.cecd.umd.edu

International Standard Book Number: 978-0-9846274-7-9

Printed in the United States of America

Printed in Southern Maryland

DEDICATED TO

DILIP ANIL “NEIL” ANAND

Preface and Acknowledgements

While the "engine" and the concept of engineering are hundreds of years old, the impact of engineering on society evolved only gradually until the Agricultural Revolution. At that point many land grant universities like ours were created to address the issues of agriculture, and a fairly common engineering curriculum was created that emphasized fundamental and very practical knowledge. The Industrial Revolution brought an explosion of inventions and innovations that led to a revision of the engineering curriculum, but it still remained within the confines of mostly fundamental and application engineering courses. Many of the technological advances were in response to government's research and development efforts during multiple wars. The resulting engineering innovations spilled into the commercial sector and had an exponential impact on society. In the last few decades, the Internet and other modern communication technologies have made the world instantaneously connected by making raw information available to almost everyone. The reach of engineering has thus become all-pervasive, and the engineer can no longer be consigned to a narrowly focused curriculum, but must become a global engineer.

In an examination of the role that engineers might play in 2020, the National Academy of Engineering noted that they "aspire to a future where engineers are prepared to adapt to changes in global forces and trends and to ethically assist the world in creating a balance in the standard of living for developing and developed countries alike."[i] Several universities (Berkeley, MIT, Carnegie Mellon, Stanford, etc.) have

---

[i] *The Engineer of 2020: Visions of Engineering in the New Century*, National Academy of Engineering, National Academies Press, 2004.

already taken steps to recognize this expanded role for engineers by introducing courses in humanitarian engineering, development engineering, and engineering and public policy.

We at the University of Maryland decided to contribute to that thought process by introducing a course entitled Engineering for Social Change. The purpose of this course is to inculcate in our undergraduate students an appreciation for the social change that engineering creates and to inspire students to practice social entrepreneurship and pursue ideas that make a difference in this new global environment. Our new mantra, "Engineering is not just Engineering," refers to the fact that much engineering work requires a significant percentage of "soft" skills as well as the traditional technical skills. We also make the point that many types of global, social problems cannot be solved by using technical skills only; rather, they require a new approach. In the course, we discuss various examples of catalysts and game changers for social change in different engineering activities and allow the students to engage in social entrepreneurship. At the culmination of the semester, the students award $10,000 to a deserving nonprofit organization engaged in effective social change.

Our course was offered in collaboration with the School of Public Policy for two semesters to a total of 72 engineering students, each one of whom contributed a blog post at the end of the semester. As part of the entrepreneurship experiment, the students chose to give $10,000 for the construction of a vegetable garden for the education of young children and $10,000 for the construction of a well to provide fresh water to thousands of students in Sierra Leone. Our thanks go to the Neilom Foundation for providing the grants to support this effort. Although all the students in our class stayed highly engaged and had positive comments, their overall impression is best captured by one of our students, Julian Lofton.

> *Giving money to charity is easy; inducing social change is much more difficult, but the impact is much greater. As engineers, we are given the opportunity to do much more with our work. Our solutions can alter the way we live, the way we view ourselves, and the way we view one another. How we use this opportunity is up to us.*

The purpose of writing this book was two-fold. The first was to document why we felt obligated to introduce a course in Engineering for Social Change and to demonstrate how we see the developed and developing world accept technology. The second reason was to give our students a kaleidoscopic view of engineering and some of its consequences – how the world has become instantaneously connected and the role that they have to play is larger than what they thought, and how engineering is not just engineering, but much more. This view is best captured by Douglas, Papadopolous, and Boutelle:

> *The blending of engineering and citizenship is nothing new. What's new is that engineers are being asked to extend their sphere of responsibility into new areas: the developing world, the environment, the proper use of intellectual property, security and privacy issues; the list goes on. At the same time, society is asking engineers to accept more responsibility for the impacts of the products and services they design. The world is not blaming engineers for climate change, loss of data privacy, and so on, but society is making it overwhelmingly clear that since engineers had a role in creating these challenges, engineers must accept the role in addressing them as well.*[ii]

As with any such endeavor, we are thankful to many colleagues and friends who gave us advice and engaged in lengthy discussions. The whole project started with our discussions with School of Public Policy Professor Robert Grimm and then with Associate Director Jennifer Littlefield. They became partners in teaching the course. Emeritus Dean George Dieter and ME Chair and Professor Balakumar Balachandran were early and consistent supporters. Professor Jungho Kim was on board from the very beginning and happily gave a lecture each semester. Additional Invited Lectures were given by University of Maryland Professors Mark Fuge, Millard Firebaugh, Monifa Vaughn-Cooke, Barton Forman, Steven Gabriel, Jelena Srebric, and Richard McCuen. Lectures were also given by Professor Andrew Miller of the University of Illinois at Urbana-Champaign, Erica Estrada-Liou from the Academy of Innovation and Entrepreneurship, Ryan Shelby from USAID, Cam Pascual of the Food Recovery Network, Sagar Doshi of Deloitte, Kevin Cleary of the Sheikh Zayed Institute for Pediatric Surgical Innovation, Smeeta Hirani of World Pulse, and Suchita Guntakatta from the Bill and Melinda Gates Foundation. Terry Island of the Undergraduate Office helped gather our first 32 students. Our undergraduate helpers for the course were Carolyn Plitt and Ruth Baldwin. Our graduate student Abrar Saleem Mohammed helped in preparing the Appendix and formatting the

---

[ii] Douglas, D., Papadopolous, G. and Boutelle, J., *Citizen Engineer: A Handbook for Socially Responsible Engineering*, Pearson Education, 2009, p. 6.

references. Staff member Ania Picard helped in the production of the book. Emeritus Dean George Dieter thoroughly reviewed the entire manuscript. Lise Crittenden, past CECD staff member and now in South Carolina, edited the book. Other reviewers also included Cheryl Wurzbacher from CALCE – University of Maryland and Ashwin D'Cruz from the University of Western Australia. As in past works, Eric Hazell did the final editing. Finally, and perhaps most importantly, thanks to our 72 students who displayed much enthusiasm for the course and for giving away money for a good cause.

The Authors

College Park, Maryland
July 2016

Foreword

When I was first told by the authors that they were putting this unique book together on the interplay between engineering and social change, I welcomed their bold initiative. Now, I am delighted to see this rather unconventional book in print. I believe this terrific endeavor to inform our students and help them learn about the manner in which one might use engineering knowledge and skills to impact social changes is quite timely and will continue to remain relevant in our globally inter-connected landscape. We are lucky indeed to have this well-conceived book written by educators and researchers, who collectively between them have many decades of experience. Their tremendous wealth of experience with curriculum, inter-disciplinary pursuits honed by work in world class research centers, and global perspectives have enriched this book and provided some fresh avenues for an engineer to explore his/her passion and appreciate how technology impacts society and vice-versa. I believe that this book can be an excellent catalyst to further curricular efforts related to engineering and social change.

I can heartily recommend this book to all who would want to see a broad-ranging treatment that brings new insights to an important, emerging field.

Balakumar Balachandran
Chair and Minta Martin Professor
Department of Mechanical Engineering
University of Maryland, College Park, MD
July 20, 2016.

Foreword

Engineering for Social Change is a unique book. There are few books devoted to showing how engineering directly and indirectly affects society. Written by faculty and staff of Mechanical Engineering in the Clark School of Engineering at the University of Maryland, the first two-thirds is devoted to describing how the work of engineers is rapidly changing the world around them, and how these changes will affect the careers of future engineers. The authors draw on a vast array of developments and ideas from the Internet to support their positions. The last third of the book describes in detail a course developed for senior students in Mechanical Engineering to open their eyes to and embrace these changes. In addition, the course introduces them to the concept of philanthropy and guides them through the decision process of actually spending real money on a worthwhile social improvement project.

George E. Dieter, ScD.
Member of the National Academy of Engineering
Emeritus Dean, Clark School of Engineering

Author Biographies

Davinder K. Anand is Professor Emeritus of Mechanical Engineering and Director of the Center for Engineering Concepts Development, both at the University of Maryland, College Park. He received his doctorate from George Washington University in 1965. Dr. Anand was Senior Staff at The Applied Physics Laboratory of the Johns Hopkins University from 1965-1974. From 1991-2002, he chaired the Department of Mechanical Engineering at College Park. He has served as a Director of the Mechanical Systems Program at the National Science Foundation, and his research has been supported by NIH, NASA, DOE, DOD, and industry. He has lectured internationally, founded two high technology research companies (TPI Inc. and Iktara and Associates, LLC), published eleven books and over one hundred and seventy papers, and has one patent. He is a Distinguished Alumnus of George Washington University, and was awarded the Outstanding and Superior Performance Award by the National Science Foundation. Dr. Anand is a Fellow of ASME and is listed in *Who's Who in Engineering*. He is also the founder and Executive Director of the Neilom Foundation, a Maryland nonprofit working to enrich the lives of young people. He is a member of the Cosmos Club.

Dylan A. Hazelwood is the Assistant Director of the Center for Engineering Concepts Development, in the Department of Mechanical Engineering at the University of Maryland. He received a Bachelor's Degree of Applied Computing from the University of Tasmania, Australia. He has expertise in information technology systems and development. He headed up the information technology group for the Department of Mechanical Engineering and was involved in information technology infrastructure development and management, high performance computing cluster development and implementation as well as establishing distance learning and other educational technologies. He also worked with the Energetics Technology Center in Southern Maryland in the areas of informatics and IT management. Since joining CECD he has continued to work on energetics informatics, rare earth materials research and STEM program analysis. He co-authored the 2012 book *Rare Earth Materials: Insights and Concerns* and the 2013 book *S&T Revitalization: A New Look*. He has been the course manager for CECD's Engineering for Social Change course since its inception, and in 2016 spearheaded an effort with the College of Southern Maryland to support a successful pilot program of a student-led social entrepreneurship course in the Business and Management Division.

Michael G. Pecht is a world-renowned expert in design, test, and risk assessment of electronics and information systems. Prof. Pecht has a BS in Physics, an MS in Electrical Engineering and an MS and PhD in Engineering Mechanics from the University of Wisconsin at Madison. He is a Professional Engineer, an IEEE Fellow, an ASME Fellow, an SAE Fellow and an IMAPS Fellow. He is the editor-in-chief of IEEE Access, and served as chief editor of the IEEE Transactions on Reliability for nine years and chief editor for Microelectronics Reliability for sixteen years. He has also served on three US National Academy of Science studies, two US Congressional investigations in automotive safety, and as an expert to the US Food and Drug Administration (FDA). He is the founder and Director of CALCE (Center for Advanced Life Cycle Engineering) at the University of Maryland, which is funded by over 150 of the world's leading electronics companies at more than US$6M/year. The CALCE Center received the NSF Innovation Award in 2009 and the National Defense Industries Association Award. Prof Pecht is currently a Chaired Professor in Mechanical Engineering and a Professor in Applied Mathematics, Statistics and Scientific Computation at the University of Maryland. He has written more than twenty books on product reliability, development, use and supply chain management. He has also written a series of books of the electronics industry in China, Korea, Japan and India. He has written over 700 technical articles and has 8 patents. In 2015 he was awarded the IEEE Components, Packaging, and Manufacturing Award for visionary leadership in the development of physics-of-failure-based and prognostics-based approaches to electronic packaging reliability. He was also awarded the Chinese Academy of Sciences President's International Fellowship. In 2013, he was awarded the University of Wisconsin-Madison's College of Engineering Distinguished Achievement Award. In 2011, he received the University of Maryland's Innovation Award for his new concepts in risk management. In 2010, he received the IEEE Exceptional Technical Achievement Award for his innovations in the area of prognostics and systems health management. In 2008, he was awarded the highest reliability honor, the IEEE Reliability Society's Lifetime Achievement Award.

Mukes Kapilashrami will be joining the United Nations as Programme Management Officer of the Division of Environmental Affairs. Currently he is a Senior Research Associate with the Center for Engineering Concepts Development, at University of Maryland and is also an Affiliated Scientist at the Advanced Light Source at Lawrence Berkeley National Laboratory. He obtained his doctorate degree in experimental materials physics from the Royal Institute of Technology in 2009, and has held postdoctoral appointments at Uppsala University, Lawrence Berkeley National Laboratory and University of Wisconsin–Madison. Dr. Kapilashrami's research interests include understanding and tailoring the intrinsic properties of semiconductor materials complexes at nanoscale for application in alternative green energy technologies. He is an internationally recognized scientist, widely published (400+ citations), and an active member of several international scientific review panels as well as a nominated mentor by the US Department of Energy Office of Science's Research Internship Program. With nearly 10 years of international non-profit fieldwork experience in empowerment of underprivileged youth through education, Dr. Kapilashrami has been an advocate of industry and academia working in concert to exploit science and technology in solving the most pressing issues facing humanity.

Acronyms

ABET	Accreditation Board for Engineering and Technology, Inc.
AI	Artificial Intelligence
APL	Applied Physics Laboratory
AR	Augmented Reality
ASCE	American Society of Civil Engineers
ASME	American Society of Mechanical Engineers
B-CORP	Benefit Corporation
BS	Bachelor of Science
CALCE	Center for Advanced Life Cycle Engineering
CDD	Cooling Degree Days
CECD	Center for Engineering Concepts Development
CFD	Computational Fluid Dynamics
COP21	Conference of the Parties
COPD	Chronic Obstructive Pulmonary Disease
CPNL	Center for Philanthropy and Non-profit Leadership
CPU	Central Processing Unit
CSM	College of Southern Maryland
CSR	Corporate Social Responsibility
CT	Computed Tomography
DDT	Dichlorodiphenyltrichloroethane
DEHP	Di(2-ethylhexyl) Phthalate
DHCD	Department of Housing and Community Development
DiNP	Diisononyl Phthalate
DNA	Deoxyribonucleic Acid
DoD	United States Department of Defense
DOE	Department of Energy
DOJ	Department of Justice
E4C	Engineering For Change
EMDC	Emerging Markets and Developing Countries
ENSO	El Nino Southern Oscillation

EPA	Environmental Protection Agency
EWB	Engineers Without Borders
FBI	Federal Bureau of Investigation
FDA	Food and Drug Administration
GDP	Gross Domestic Product
GE	General Electric
GEF	Global Environment Facility
GM	General Motors
HFT	High Frequency Trading
HIPAA	Health Insurance Portability and Accountability Act of 1996
HIV	Human Immunodeficiency Virus
HTC	Taiwanese Consumer Electronics Company
H&M	Hennes & Mauritz AB
IC3	Internet Crime Complaint Center
IEEE	Institute of Electrical and Electronics Engineers
IMF	International Monetary Fund
IOT	Internet of Things
ISCC	Ideas for Social Change Challenge
ISIS	Islamic State of Iraq and Syria
IT	Information Technology
JHU	Johns Hopkins University
KPMG	Professional Services Company
LED	Light Emitting Diode
LI-FI	Light Fidelity
LIDAR	Light Imaging, Detection, And Ranging
LINC	Laboratory Innovation Crowdsourcing
M-PESA	Mobile Phone-based Money Transfer and Financing
MEMS	Microelectromechanical Systems
MIT	Massachusetts Institute of Technology
MOOC	Massive Online Open Course
MRI	Magnetic Resonance Imaging
NAE	National Academy of Engineering
NAS	National Academy of Sciences
NASA	National Aeronautics and Space Administration
NASDAQ	National Association of Securities Dealers Automated Quotations
NGO	Non-Governmental Organization
NIH	National Institutes of Health
NIST	National Institute of Standards and Technology
NOAA	National Oceanic and Atmospheric Administration
NRA	National Rifle Association
NRDC	Natural Resources Defense Council

NSF	National Science Foundation
NSPE	National Society of Professional Engineers
NYSE	New York Stock Exchange
OECD	Organization for Economic Cooperation and Development
OER	Open Educational Resources
PCBs	Polychlorinated Biphenyls
PE	Professional Engineer
RFID	Radio Frequency Identification
RFP	Request For Proposal
RUS	Department of Agriculture Rural Utilities Service
SoC	System-on-a-Chip
SPDI	Solar Powered Drip Irrigation
STAR	Smart Tissue Automated Robot
STEM	Science, Technology, Engineering and Mathematics
SWIFT	Society for Worldwide Interbank Financial Telecommunication
TL:DR	Too Long: Didn't Read
UAE	United Arab Emirates
UAV	Unmanned Aerial Vehicle
UCLA	University of California, Los Angeles
UN	United Nations
UNDP	United Nations Development Programme
UNGEI	United Nations Girls' Education Initiative
USAID	United States Agency for International Development
USDA	United States Department of Agriculture
USGS	United States Geological Survey
VNC	Virtual Nonprofit Challenge
VR	Virtual Reality
WHO	World Health Organization
WI-FI	IEEE 802.11x Wireless Standard
WTO	World Trade Organization

Contents

This page intentionally left blank.

Traditional Chinese rickshaws pass by a lamp oil company [1]

A modern Tesla showroom, showing the company's electric Model S [2]

Chapter 1

The Changing Face of Engineering

Toto, I've a feeling we're not in Kansas anymore.[3]

The National Academy of Engineering (NAE) predicts that "the pace of technological innovation will continue to be rapid (most likely accelerating). The world in which technology will be deployed will be intensely globally interconnected."[4] The nature of this rapid technology growth gives rise to myriad complex and nontraditional issues that engineers are being called to engage. The professional practice of engineering, its impact on society and resulting level of responsibility of the engineer, is undergoing significant change, and engineering education must take that into account.

A bright and successful future for mankind calls for a new kind of engineer to solve the world's most pressing problems, one who will move from a focus on developing consumer-driven technological solutions for the world's wealthiest, to balancing that with greater development of innovative and low cost solutions directly for the needs of the developing world. As the gap widens between the rich and the poor throughout the world, engineers must accept and apply novel approaches to understanding cultural forces. This is also of pressing importance as societies, both developing and developed, face the unintended consequences of tremendous past engineering successes. These are in areas such as waste, sustainability, and the economy, but also in the misuse of technology. Many of these issues have clear policy implications, requiring more engineers to understand and engage in public policy. In the ensuing paragraphs we will elucidate various salient points in some selected areas to develop a keener understanding of the new knowledge base required for the engineer of the future.

1.1 A Changing World

Many of the 7.4 billion people of our world are simply not able to keep pace. While technology is having an increasing impact on our daily lives in the developed world, many of the benefits to our society are not impacting developing countries. When hailing an Uber on your smartphone or ordering fast food online, consider that over half of the 700 million people living in Africa live on less than $1 a day, continuously facing malnutrition, dehydration, and disease.[5] While some parts of Africa may be at the extremes, millions of people in the rest of the world live with very little earnings and expend much of their energy and time simply in survival, rather than enriching their communities through education to participate in the new global economy.

Global poverty signifies a growing need for a refocus of engineering on the world's poor, instead of on supplying more technology to the world's rich as the gap between those with and without technology is widening. Many problems in the developed world are well addressed by creative and effective technological solutions, and engineers have moved on to new challenges. However, while these solutions are commonplace in developed countries, in the developing world people struggle to even afford the most basic of technologies such as mobile phones. "Around the world right now there are more people living in darkness than when the light bulb was invented. For those lucky enough in the developing world to have kerosene lamps, a million will die due to those same lamps."[6]

In order to address these problems, engineers should consider developing radical, low-cost solutions. For instance, in July 2015, a company named d.Light successfully optimized product function, simplicity and affordability when it launched a $5 solar-rechargeable LED lantern for sale in the developing world.[7] This unit is able to provide light for many hours a day to provide greater safety and to enable opportunities for education and greater overall self-sufficiency for millions of people without a supply of electricity.

In late 2015, the United Nations defined seventeen global Sustainable Development Goals under the 2030 Agenda for Sustainable Development, with specific targets for the fifteen-year period. As clearly seen, engineering plays a significant role in addressing almost every goal, as follows:[8]

- End poverty in all its forms by 2030.
- End hunger, achieve food security and improved nutrition and promote sustainable agriculture.
- Ensure healthy lives and promote well-being for all at all ages.
- Ensure inclusive and quality education for all and promote lifelong learning.
- Achieve gender equality and empower all women and girls.
- Ensure access to water and sanitation for all.

- Ensure access to affordable, reliable, sustainable and modern energy for all.
- Promote inclusive and sustainable economic growth, employment and decent work for all.
- Build resilient infrastructure, promote sustainable industrialization and foster innovation.
- Reduce inequality within and among countries.
- Make cities inclusive, safe, resilient and sustainable.
- Ensure sustainable consumption and production patterns.
- Take urgent action to combat climate change and its impacts.
- Conserve and sustainably use the oceans, seas and marine resources.
- Sustainably manage forests, combat desertification, halt and reverse land degradation, and halt biodiversity loss.
- Promote just, peaceful and inclusive societies.
- Revitalize the global partnership for sustainable development.

It is important to note that low cost and low complexity solutions to the world's most challenging problems do not preclude the profit motive for companies. Companies in the developed world are now seizing opportunities to tackle social issues whilst also making a profit from doing so. This is a key growth area for engineers, as, according to management theorist Michael Porter, businesses are now looking to embrace large-scale opportunities for social impact where they can build profits at the same time.

> *The ultimate impact businesses can have is through the business itself....There are huge unmet needs in the world today. The question now is how to get capitalism to operate at its best because capitalism is fundamentally the best way to meet needs. If you can meet needs at a profit, you can scale.*[9]

With increasing manufacturing activity and the use of more raw materials, the engineering of sustainable items takes on greater meaning. In the *Citizen Engineer*, the authors describe the challenges of engineering a 60-watt light bulb for one billion people. The total weight of the bulb and packaging is 20,000 metric tons. Assuming they are used for four hours a day, this consumes a combined 10,000 megawatts of electricity. This would require the building of 20 new 500-megawatt power plants that are estimated to burn almost one and a half million tons of coal every year. If renewable energy were utilized instead, this would require the installation of 50-square-kilometers of solar panels. The scale of the materials and energy used in the lifecycle of that light bulb are also very significant. These include the materials for their construction, energy used in the factories for production, materials for packaging, fuel in shipping and land transportation and the space in landfills that the bulbs will take up when no longer working. The authors further note, "... if we're having trouble delivering a single light bulb sustainably, what happens when these billion people want stoves, refrigerators, TVs, computers, cell phones, radios, and cars?"[10]

> *We now have the ability to produce things by the billions, the capacity to distribute them globally, and the markets to consume that kind of scale. Global markets, global fashions, and global consumer trends result in mass production of successful products and the repercussions of success can far exceed anything engineers originally envisioned....did Henry Ford have any inkling that a hundred years after the introduction of the Model T we'd be trading streets full of manure for global climate change, due in part to the exhaust of more than a billion cars?*[11]

As computational power has grown over time, the scale of materials used by engineers now varies from subatomic to many tons of material. Beyond physical dimensions, the materials engineers currently use may have an impact many centuries from now. An aluminum can may take as long as 200 years to break down in the environment, while plastic bottles take 450 years, and fishing line takes an estimated 600 years. High-level waste from nuclear power plants in the form of Plutonium-239 has a half-life of 24,000 years. Glass bottles are estimated to take a staggering one million years to break down.[12] To say that the effects of engineering on the world will have a lasting global impact is an understatement of extraordinary measure.

With new materials, factories, suppliers and customers spread throughout a new global marketplace, engineers now have greater reach. Teams of engineers work together across different continents and time zones on solutions for application in many different markets. Research, development and manufacturing of all types of products and components are outsourced to countries around the globe. Engineers now work on the development of products and services at scales far greater than ever before, with the possibility of hundreds of millions or even billions of users worldwide, with a tremendous total social impact.

Industries today are working around the clock to meet production demands of our technology-driven society. Combined with the development of communication technologies, this has opened the doors for local and even global suppliers and vendors to support larger producers to help them meet their delivery deadlines. Modern and lean manufacturing combined with improvement in planning logistics and supply chain optimization will continue to have an impact on how and where items are made. Lower local inventory and a worldwide supply chain are requirements. The average American car is now made with components from around the world. According to a 2015 analysis by *Consumer Reports*, the most "Made in the USA" car was made from only 75% American-sourced parts, and this definition includes Canada and Mexico. We will never be able to manufacture a car (and many other products) that are "made 100 percent in one country anymore ... What you'll typically see instead is larger components made near the point of sale, to save shipping costs, while small components like electric motors and actuators will be brought in from anywhere."[13]

Globalization and the industrialization of developing countries have resulted in systematic outsourcing of manufacturing of goods from developed countries to take advantage of lower labor costs. However, this has reduced employment opportunities in the manufacturing sector in economically developed countries such as the US due to higher wages and greater worker protections.[14] This is irreversible and inevitable, requiring the reorganization of the labor markets in the developed countries. The developed world can no longer expect that we can export our problems to other parts of the world. These include shipping electronic waste internationally, burying trash in endless landfills, and using materials and labor in countries with lax laws, all of which result in environmental and social damage, etc. Engineers now deal with the global societal and environmental consequences of contemporary engineering and will need to not only face these overwhelming challenges but also continue to improve the quality of and access to food, clean water, shelter, medicine, and clean air everywhere in the world.

1.2 The Impact of Society on Technology

While much of this book is focused on the impact of engineering on society, it is important to also consider the impact of society on the development and adoption of engineered technologies.

> *Modern technology is not simply the rational product of scientists and engineers that it is often advertised to be. Look closely at any technology today, from the aircraft*

> *to the Internet, and you'll find that it truly makes sense only when seen as part of the society in which it grew up.*[15]

To ensure success, engineers must create their designs in harmony with community members, and with local social norms and practices in mind. Often the differences between success and failure are not engineering factors. Henry Petroski tells the story of what pencil designers did in the late 1800s:

> *In order to get around the growing shortage of red cedar, [they] devised a pencil with a paper wrapping in place of the normal wood. It worked well technically and showed great promise but the product failed for unanticipated psychological reasons. The public, accustomed to sharpening pencils with a knife, wanted something that could be whittled. The paper pencil never caught on.*[16]

The "green explosion" of the past decade has seen the growing importance of recycling, reducing energy usage, and increasing sustainability of products and services. Sustainable engineering is now a selling feature in many common products. These include thinner water bottles using less plastic, products incorporating recycled materials, items that use sustainably sourced, and natural, renewable materials such as bamboo. Not only should materials used to create products preferably be renewable, safe, and "green," but they should also be sourced from companies that operate responsibly. This includes decreasing pollution, treating employees fairly, and engaging in business activities that do not exploit others. For instance, in July 2011, Motorola Solutions, Inc. and AVX announced the Solutions for Hope Project (Solutions for Hope 2015). This project is focused on sourcing conflict-free tantalum from the Democratic Republic of Congo and promoting the economic stability of the area.[17] In 2012, Intel followed by setting a goal across its complete supply chain to create its microprocessors using metals (tungsten, tin, gold, and tantalum) that were sourced from smelters not using materials from mines that were connected to human rights disputes and armed conflict.[18] As another example, Apple recently purchased more than 36,000 acres of forestland to preserve it and sustainably source paper for their future products.[19]

Corporate social responsibility (CSR, also referred to as corporate conscience or corporate citizenship) is a key consideration that plays an important part of the perception of a company and its products. A

Neilson Global Corporate Citizenship Survey found that almost half of global consumers were willing to pay more for products and services from a company that exhibited evidence of projects that helped those in need.[20] Some companies have now embedded social responsibility into the core of their business, with the mantra now being that demonstrating social responsibility is good for business. This takes many forms when selling products or services, such as:

- One-for-one, where a customer buys a product with the knowledge that the company will then give another of the product to charity.
- Profit percentages, where a certain percentage of the profit made from a product is donated directly to a charity, sometimes selected by the customer.
- Using profit to support charitable initiatives, local support initiatives, and free products or services to partner organizations doing social good.
- One-for-other, where customers are encouraged to donate items to a charity in exchange for a discount or special items from a company, or where customers buy a product and the company reciprocates by supporting charitable activities in a specific amount (i.e., one month of clean water for a child).

On the company side, these actions may include (among many others):

- Philanthropic activities by employees, supported with company time and funds, and foundations formed in the name of the company to engage in social good. There are many engineering firms with their own foundations, some spending millions of dollars each year to support charitable activities.
- Corporate Social Responsibility transparency and reporting, such as Hershey's annual CSR report.[21] This covers the work the company is doing to benefit society, such as what they have achieved each year to improve the sustainability and transparency of cocoa and palm oil growers and suppliers, and their efforts to support training thousands of farmers.
- Reducing the use of dangerous or environmentally insensitive materials or production processes. Buying carbon credits or being "carbon-neutral" is also a popular choice to offset manufacturing pollution.

- Donation of time and talent in pro bono projects for non-profit organizations or direct education and support of local community members.

Culture also plays a role in defining how engineers work and the products and services they create. For a discussion on the rules of culture in the creation and adoption of technology, see Chapter 2.

1.3 A New Kind of Engineer

Engineers for generations have invested their time and energies into creating products that had the possibility of making them wealthy and internationally renowned if ultimately successful. Profit will always remain a clear motive for technological development by engineering corporations, and while these companies have been very successful, many have tuned into the changing views held by modern society and are getting involved in activities for doing social good. As a result, creating a positive social impact with a business's product or service is becoming another contributing factor in being successful.[22] With 24.9% of people in the US over the age of 16 volunteering their time to help others,[23] evidence shows that many younger engineers are interested in not only being involved in opportunities to give back during their education, but want to also find personal fulfillment in creating positive social change throughout their careers.

The world facing the engineer of the future is one of complex and global-scale challenges. One of the side effects of our extraordinary improvements in communications technology is that we now can no longer ignore challenges that exist outside of our communities. To address these challenges, the call from many sources is for a new type of engineer, a multi-talented person capable of wielding formidable technical skills but also of applying those skills to our new world in a social and political context, as well as maintain a truly global perspective. An engineer is no longer simply the technical problem solver in a local environment. Products and services are created in teams around the clock for use in locations across the world. Policy makers will turn to engineers to create the technologies of tomorrow to solve worldwide issues of poverty, clean water and sanitation, reliable clean energy and inequality. Not only do engineers need to deliver a technology-rich future, but at the same time they must save us from ourselves and help to repair the damage done by unrestrained growth, preserving a future for generations to come.

> *As a society we have never before faced a problem like human-induced climate change. Never before have we had the capacity to produce and share so much data about our world, our lives and our finances. Never before have had we had a human population of 7 billion people, all needing food, water, shelter, education, employment and healthcare. In a competitive global market, engineers must constantly innovate to create new solutions and invent new ways of solving problems. Engineers who expect to provide the same, standard answers in an ever changing, complex world, will soon be out of work.*[24]

A traditional, technically focused engineer is no longer sufficient. Engineers of the future will need to utilize their technical prowess along with social awareness and sustainable development skills, working with a wide variety of stakeholders globally to address these challenges. While more respected in other countries[i], the engineer in the US has been traditionally viewed as much lower than other professions such as law and medicine, and their starting salaries are typically far less. Even in their role as the architects of our modern human existence around the world, engineers have not enjoyed high esteem in the US. A Harris poll from the American Association of Engineering Societies notes that the public in the US is currently disengaged from a clear understanding of the role of engineers in society:

> *...the survey showed that only 28 percent of people believe that engineers are sensitive to societal concerns; only 22 percent thought engineers improve people's quality of life; and only 14 percent of people believe that engineers save lives.*[25]

The moral and social responsibility in the practice of engineering is a shared burden resting not just upon the shoulders of the engineer, but also upon a number of other entities, including governments, non-government organizations (NGOs), nonprofits, industry, professional societies, and universities. Ursula Franklin, a metallurgist and Professor Emeritus at the University of Toronto, espouses a mindset of "justice, fairness and equality in the global sense." Her recommendations for

---

[i] In Panama, for example, engineers are addressed as "Engineer LASTNAME" and are seen as equivalent to doctors.

engineers are to ask seven questions of an engineering project. Does the project:[26]

- Promote justice?
- Restore reciprocity?
- Confer divisible or indivisible benefits?
- Favor people over machines?
- Minimize or maximize disaster?
- Promote conservation over waste?
- Favor reversible over irreversible?

A reasonable question to ask is, why engineers should care beyond a shared concern for the world they live in, particularly those who are just entering the profession, and may be working on projects where their input is minimal? A surprising statistic shows that the most common undergraduate degree of Fortune 500 company CEOs is engineering.[27] Many engineers will at some point during their career move to management positions, where their influence over projects that directly affect society will increase. When they do have a say over the direction a product or service might take, they can choose to undertake careful analysis to make a decision with a positive social impact, rather than ignoring that impact.

There are multiple ways in which positive social change can be achieved throughout the engineer's career. The first is via the creation of a well-designed and successful product or service that is adopted by many millions of people. This results in profit for the company and simultaneously changes the way we conduct our daily lives. An example of this is Apple or Samsung with their extraordinarily popular smartphones. These companies do not necessarily have positive social change as a mantra, but it is a byproduct of the products and services they create and sell. Another method is to develop products and services specifically with positive social change as a goal. Many companies who do this have chosen to be certified as B-Corps, or benefit corporations. These are for-profit, legal business entities, where the corporation's goals are not only to make a profit, but also to positively impact the environment and society. They are required under this classification to publish annual public transparency reports showing what they have undertaken to help society as part of their classification. Examples include:

- d.light design, a global company developing extremely low cost solar lighting solutions for use in the developing world[28]

- PV Squared, a worker-owned solar installation company promoting solar power for sustainable communities[29]

Finally, engineers can create positive social change through nonprofits and NGOs. These organizations are created purely with positive social change in mind, generally demonstrate the greatest transparency, and display a clearly stated focus on working towards creating a positive impact on society in well-defined areas. Examples include the Bill and Melinda Gates Foundation, which works on a variety of issues to improve lives in the developing world, and Charity:Water, which provides clean and safe drinking water. Perhaps the most visible nonprofit engaging engineers in positive social change in the US and abroad is Engineers Without Borders (EWB), which is deeply integrated into the undergraduate engineering programs, often under the guidance of an experienced engineering professional. Engineering students, faculty, and professional engineers can volunteer their time and skills to work together on EWB projects. EWB also offers the services of their most skilled volunteers via the Engineering Service Corps to organizations involved in international development activities.[30] Another large collection of engineers with a development engineering and social change mindset is Engineering For Change (E4C), which is an online community of 20,000+ members. The organization provides an online academy, an innovation laboratory and a media platform that all demonstrate the engineering challenges and solutions found in the developing world.

1.4 Engineering Education

Engineering education has a moral and social responsibility for providing the right type of opportunities for engineers to learn technical and soft skills, empowering them to effectively work in a global environment. The engineer must also be aware of public policy, be willing to develop a global perspective, be technically skilled, and be ready to address the great challenges facing society in the future. But the question is: How effective are we in preparing our future engineers?

A 2013 study by Rice University Professor Erin Cech discovered that engineering students she polled throughout their four years became more cynical by the end of their engineering undergraduate degree as a result of a "culture of disengagement" fostered at each of the four Northeast US universities in her study. Cech tracked four areas of importance: professional and ethical responsibilities, an understanding of the consequences of technology, social consciousness and an understanding of how people use machines. Her perspective was that:

> *Issues that are nontechnical in nature are often perceived as irrelevant to the problem-solving process ... There seems to be very little time or space in engineering curricula for nontechnical conversations about how particular designs may reproduce inequality—for example, debating whether to make a computer faster, more technologically savvy and expensive versus making it less sophisticated and more accessible for customers.*[31]

The number of credits required for the baccalaureate degree has also been reduced over time, moving from 144 down to 124 over the last 50 years, so even less space is available for non-technical classes. Paul E. Jacobs, Qualcomm CEO, notes that:

> *....today's engineering education seems stuck in the past. Professors often stand in front of students giving dry lectures based on notes they wrote years ago. However, students are now accustomed to quickly obtaining the information they need through rich multimedia content on the Internet.*[32]

The traditional approach in engineering education had a singular focus on transferring as much technical content as possible to the student through regular "book courses," where a textbook is carefully followed and the content is very easily defined within the specific subject. Less effort was expended in creating a holistic engineer, one with more than just technical skills. While many years ago engineers were able to create products and services without many of the modern policies now in place that guide the practice, the modern engineer must learn to create and operate with a wider perspective. But is there room for this approach in a traditional degree program?

> *Among the principal professions, engineering is the only one for which the bachelor's degree is the primary accredited, professional degree....since the entire professional program is concentrated into the undergraduate degree, engineering education has little room, if any, for much needed breadth.*[33]

The teaching of engineering design has generally not considered in depth the societal ramifications of how a product may be used or misused, how a technology may put people out of work, how it may impact the environment beyond its functional life, or many of the long-tail unintended consequences of various engineered products and systems. The NAE asks the following provocative questions:

> *Should the engineering profession anticipate needed advances and prepare for a future where it will provide more benefit to humankind? Likewise, should engineering education evolve to do the same?*[34]
>
> *Just as important will be the imperative to expand the engineering design space such that the impacts of social systems and their associated constraints are afforded as much attention as economic, legal, and political constraints.*[35]

In order to address larger social issues, we need to reverse this situation. By the time engineering students finish their degree, they must be endowed with, and convinced that they have the skills to, make an impact and change the world for the better. While some universities have developed forward-looking and nontraditional approaches to develop the skills required, but those that offer a more traditional engineering education do not appear to have been widely addressing this need. Debbie Chachra notes that engineering is typically being taught with a lack of humanization. She states that "students march through multiple science and mathematics courses before they begin to 'do' engineering, and almost all of the engineering work they do is analytical, focused on the right answer, and devoid of the ambiguity associated with working with real people."[36]

Partially due to the amount of technical content required in undergraduate engineering education, there is a lack of scope in engineering degrees. The technical focus does not allow for a greater understanding of public policy, and there is much more a focus on hard skills rather than 'soft skills.' The NAE has noted the challenges of producing graduates who are technically strong engineers but who often "lack the people skills (also called 'soft' skills) that enable them to meet their full potential. Today's engineers need to be not only technically strong but also creative and able to work well in teams, communicate effectively, and create products that are useful in the 'real world.'"[37]

Without effective public policy, the creations of engineers may be misused, may never make it into the hands of those who might use them, or may be restricted such that they are ineffective at best. Engineers have a significant role to play in not only understanding public policy ramifications for the products and services they are developing, but also to inform and educate both those who make policy - and members of society who help influence policy - so that they truly understand the reasons for creating policy.

> *... engineers are also beginning to realize how important it can be to educate those who make public policy decisions about technology. Nuclear power and genetically engineered foods provide cautionary tales about what can happen when politicians and the public don't understand the realities behind new innovations.*[38]

The Institute of Electrical and Electronics Engineers (IEEE) bill themselves as "the world's largest technical professional organization dedicated to advancing technology for the benefit of humanity," and have programs at the national, regional and global levels to educate policy-makers on the social implications of technology and engage members in advocating for technology-related public policies to advance the public good.[39] Engineers must be involved in public policy, as all policy needs to be founded on solid technical ground – something most lawmakers are not well versed in, particularly with new and complex breakthrough technologies. We should heed the advice of Norman Augustine, who says that "Many of the top, pressing issues our government faces require technical solutions and would benefit from technical input."[40]

We conclude the introductory chapter with the observation that for many fundamental engineering content courses, there is a reliance on tradition and structure. Content is clear, progressive, and created with examples that illustrate the concepts learned in the main text. And yet this content often describes a fictional world in which everything works exactly as it should, with clear right and wrong answers. The real world, as we know, is very often not as simple and well-defined as a textbook example, particularly as engineering students consider how they might solve society's real and pressing social problems that require the application of engineering, public policy and skills like common sense and compassion. As a result, we believe that courses such as our *Engineering for Social Change* at the University of Maryland break out of the traditional mold for an undergraduate engineering course. We also believe it is important to include an experiential component, particularly where students have input into the decision-making components of the class. This approach, when combined with our philanthropic challenge, gives students ownership of a group decision with real impact. It is the mix of these types of experiences along with the development of technical knowledge that will ultimately create the type of engineers we need for the future, ready to take on the world's greatest challenges.

References

[1] Sowerby, A., "China, Miscellaneous Scenes: Chinese men pulling rickshaws down the street", RU 7263 - Arthur de Carle Sowerby Papers, Smithsonian Institution Archives, 1904-1954 and undated, accessed at https://www.flickr.com/photos/smithsonian/7454235686

[2] Grid Scheduler/Grid Engine, "Tesla Motors", Digital Image, Flickr, August 22, 2014, accessed at https://www.flickr.com/photos/opengridscheduler/16070358203

[3] Dorothy in *The Wizard of Oz*, Film, Metro-Goldwyn-Mayer, 1939.

[4] *The Engineer of 2020: Visions of Engineering in the New Century*, National Academy of Engineering, National Academies Press, 2004.

[5] Lupton, R. D., *Toxic Charity: How Churches and Charities Hurt Those They Help, And How To Reverse It,* HarperOne, October 2012.

[6] Developing World Technologies, 2015, accessed at http://www.developingworldtechnologies.com

[7] "D.Light launches $5 Solar Latern", Shell Foundation, July 20, 2016, accessed at http://www.shellfoundation.org/Our-News/News-Archive/d-light-Launches-$5-Solar-Lantern

[8] "Sustainable Development Goals", United Nations, 2016, accessed at http://www.un.org/sustainabledevelopment/

[9] Schiller, B., "The Future of Progressive Business Is Companies That Are Good, Not Just Doing Good", *Fast Company*, March 15, 2015, accessed at http://www.fastcoexist.com/3056482/the-future-of-progressive-business-is-companies-that-are-good-not-just-doing-good

[10] Douglas, D., Papadopolous, G. and Boutelle, J., *Citizen Engineer: A Handbook for Socially Responsible Engineering*, Pearson Education, 2009, p. 32.

[11] Ibid.

[12] "Time it takes for garbage to decompose in the environment", New Hampshire Department of Environmental Services, accessed at http://des.nh.gov/organization/divisions/water/wmb/coastal/trash/documents/marine_debris.pdf

[13] "What Makes a Car "American"", *Consumer Reports*, May 22, 2015, accessed at http://www.consumerreports.org/cro/magazine/2015/05/what-makes-a-car-american-made-in-the-usa/index.htm

[14] Gurbanov, S., "Declining Manufacturing Employment as a Global Social Change: Implications for Azerbaijan", *International Conference on "Policy Options for Social Market Economy: National and International Perspectives"*, Center for Economic and Social Development (CESD) & Konrad Adenauer Stiftung & Qafqaz University, April 9, 2013, April 9, 2013.

[15] Pool, R., *Beyond Engineering: How Society Shapes Technology*, Oxford: New York, p. 7., 1997.

[16] Ibid.

[17] Jameson, N. J., Song, X. and Pecht, M., "Conflict Minerals in Electronic Systems: An Overview and Critique of Legal Initiatives", *Sci Eng Ethics,* September 2015.

---

[18] "The World's Top 10 Most Innovative Companies of 2015 in Social Good", *Fast Company*, February 9, 2015, accessed at http://www.fastcompany.com/3041663/most-innovative-companies-2015/the-worlds-top-10-most-innovative-companies-of-2015-in-social

[19] "Apple buys 36000 acres of forest to create sustainable eco-friendly product packaging", AppleInsider, April 16, 2015, accessed at http://appleinsider.com/articles/15/04/16/apple-buys-36000-acres-of-forest-to-create-sustainable-eco-friendly-product-packaging

[20] Brooks, C., "Shoppers Willing to Pay More to Socially Responsible Companies", Business News Daily, March 29,2012, accessed at http://www.businessnewsdaily.com/2273-shoppers-pay-socially-responsible-companies.html

[21] Watt.A., "Hershey Issues Corporate Social Responsibility Report", Candy Industry, June 15, 2016, accessed at http://www.candyindustry.com/articles/87296-hershey-issues-corporate-social-responsibility-report

[22] Baillie, C., *Engineers within a Local and Global Society: Synthesis Lectures on Engineers, Technology and Society*, Morgan and Claypool, p. 7., 2006.

[23] "Volunteering in the United States, 2015", Bureau of Labor Statistics, February 25, 2016, accessed at http://www.bls.gov/news.release/volun.nr0.htm

[24] Lawlor, R., "Engineering in Society", Royal Society of Engineering, accessed at http://www.raeng.org.uk/publications/reports/engineering-in-society

[25] Goldberg D., Somerville, M., "A Different Kind of Diversity: The Changing Face of Engineering Education", The Huffington Post, August 24, 2012, accessed at, http://www.huffingtonpost.com/david-goldberg/engineering-education-reform-_b_1826537.html

[26] Baillie, C., *Engineers within a Local and Global Society: Synthesis Lectures on Engineers, Technology and Society*, Morgan and Claypool, 2006.

[27] Douglas, D., Papadopolous, G. and Boutelle, J., *Citizen Engineer, a Handbook for Socially Responsible Engineering*, Pearson Education, p. 17., 2009.

[28] d.light, 2016, accessed at http://www.dlight.com/

[29] Photo Valley Photo Voltaics, 2016, accessed at http://pvsquared.coop/

[30] "Volunteer with EWB-USA", Engineers Without Borders USA, 2016, accessed at http://help.ewb-usa.org/customer/en/portal/articles/1998795-volunteer-with-ewb-usa

[31] Maccaig, A., "Engineering education may diminish concern for public welfare issues, sociologists say", Rice University, November 20, 2013, accessed at http://news.rice.edu/2013/11/20/engineering-education-may-diminish-concern-for-public-welfare-issues-sociologist-says/

[32] The Engineer of the Future", *Forbes*, August 1, 2013, accessed at http://www.forbes.com/sites/skollworldforum/2013/08/01/the-engineer-of-the-future/

[33] Florman, S., *The Existential Pleasures of Engineering*, Second Edition, St. Martin's Griffin, February 1996.

---

[34] *The Engineer of 2020: Visions of Engineering in the New Century*, National Academy of Engineering, National Academies Press, p. 1., 2004.
[35] Ibid, p. 54.
[36] Goldberg D., Somerville, M., "A Different Kind of Diversity: The Changing Face of Engineering Education", Huffington Post, August 24, 2012, accessed at, http://www.huffingtonpost.com/david-goldberg/engineering-education-reform-_b_1826537.html
[37] Stephens, R., "Aligning Engineering Education and Experience to Meet the Needs of Industry and Society", *The Bridge*, Volume 43, Number 2, 2013, accessed at https://www.nae.edu/Publications/Bridge/81221/81233.aspx
[38] Douglas, D., Papadopolous, G. and Boutelle, J., *Citizen Engineer, A Handbook for Socially Responsible Engineering*, Prentice Hall, 2009, p. 10.
[39] "IEEE Global Public Policy", Institute for Electrical and Electronics Engineers, 2016, accessed at http://globalpolicy.ieee.org/
[40] "Engineers; Our Government Needs You", *Forbes*, March 28, 2012, accessed at http://www.forbes.com/sites/ieeeinsights/2012/03/28/engineers-our-government-needs-you/

We hope you're enjoying the book!

In case you're wondering, we left this page intentionally blank.

A group of young women consult smartphones at an amusement park [41]

A girl and a goat forage in an e-waste dump in Agbogbloshie, Ghana [42]

Chapter 2

Catalysts for Social Change

We have almost engineered everything....or have we?[43]

The growth of intellectual capital and accessibility to world markets in developing countries has led to increased globalization of ideas, technologies, processes and practices, labor, and products. As a result, social change within these countries has been significant and rapid, with a corollary effect in developed countries that face an expanded playing field. We define social change as any significant change over time in the cultural values and practices of a society. This change can be both positive and negative, and at times is an unintended consequence of the application of technology. The general pace of social change has accelerated in modern civilization as the principal activity of mankind has transitioned from hunter to inventor. This social change can be gradual or rapid, depending on the nature of the catalyst. Examples of modern day catalysts include technology (in banking, communication, commerce, entertainment, food production, medicine, transportation, and more), education, energy, culture, globalization, international trade, immigration, and many others. This chapter is centered on the impact of a few of these catalysts on the social fabric of the lives we live today. Further, we examine the role of philanthropy and nonprofits as they enhance the impact of these catalysts.

Climate change is the single most pressing issue that negatively impacts the quality of life worldwide and therefore presents itself as a significant challenge for the engineering profession to create lasting solutions. Key to slowing climate change, and therefore a companion issue, is the efficient use of energy and the increased production of clean renewable energy at a global scale. At the recent COP21 meeting in Paris, 195 countries adopted a universal climate agreement to limit global warming to 2 degrees centigrade and for the first time officially acknowledged that climate change is a significant problem affecting our species.[44]

Education stands alone in that it impacts many facets of life in an increasingly technological world. We discuss the impact of new technologies, as well as ways to bring education to a larger and more diverse population. Countries such as China and India are graduating engineers at an unprecedented rate, presenting a challenge to America to retain its position as a global technology leader. Even though many of them are of varying degrees of quality, they nevertheless represent potential competition. In recognizing the possibility that we could lose our competitive edge, the United States has launched a fairly popular STEM program, but whether it succeeds, or how well it succeeds, will not be known for a decade or more.

The level of adoption of newer technologies in the day-to-day life of people in various countries is intimately connected to cultural and political forces. Often technological advances are not acceptable for religious, traditional, and economic reasons. For example, even though modern plumbing can be easily and rapidly installed by modern equipment, it still has not made inroads in a significant part of the world. The same thing is true of education, where modern teaching is often shunned in favor of greater religious training, and certain textbooks are banned. And in China, access to Google Scholar, Wikipedia, Gmail, YouTube, Facebook, Twitter, and many more Internet applications, is blocked by the central government for political reasons.

The social impact of the various catalysts is positively enhanced by philanthropic activities, in which the United States is a world leader. With increasing scale and reach, engineers must be aware of the leverage that large-scale philanthropy can have in helping to create positive social change. There are 1.4 million nonprofits representing $5.1 trillion in total assets,[45] including a number of multi-billion dollar foundations such as the Bill and Melinda Gates Foundation, who are influencing social change in a number of areas through strategic technology investments.

2.1 Technology

Technological development has resulted in profound and far-reaching global social change across a wide variety of areas. In this section we will cover the impact of a few of these catalysts. These are communications and social media, agriculture and the food industry, transportation, and global commerce.

Communications and Social Media

Perhaps the most transformative development in the way humans interact – the development and widespread adoption of modern communications systems – has changed the very nature of our society. Since the first long-range communications via telegraph, improvements in engineering have resulted in faster and more efficient ways for individuals across the globe to communicate and share information. As a result, the decreased cost of long-distance communications has been a determining factor in mass adoption, with the cost of a call going from an equivalent of $82 per minute in 1928, to essentially free in 2016.[46]

The mobility of trading, ideas and people has increased over time as a direct result of improved communications. This has strengthened local and global economies and has brought many important social and economic opportunities for people everywhere. The rise of the Internet, email and smartphones has given us instant access to the world, granting to many the ability to work from home or on the road, giving teams the ability to work across thousands of miles and different time zones, and spreading human knowledge globally.

Consider the example of sub-Saharan Africa, which has the lowest global level of infrastructure investments, where access to mobile communications networks has increased significantly in the last two decades. Mobile phone users represented 1% of the population in 1999, but now represent the world's third-largest mobile market in terms of unique subscribers.[47] According to a study at the London Business School, there is a significant economic benefit to the adoption of mobile phones in developing countries. For every additional 10 mobile phones per 100 people in a developing country, its GDP rises by 0.5%.[48] Jenny C. Aker and Isaac M. Mbiti note that "while the telecommunications industry in the United States, Canada, and Europe invested in landlines before moving to mobile phone networks, the mobile phone has effectively leapfrogged the landline in Africa."[49] This is also true in many areas in India, China, and other developing regions.

Since the mass adoption of the Internet, we have seen improved and more sophisticated information sharing technologies. This has created a new form of media that not only has connected us more intimately, regardless of our social status and geographical location, but has also created an entirely new industry relying on the digitization, correlation, and distribution of information.

> *There is extensive computerization of all organized activities as well as exponential growth in the production and flow of texts, images, and data by the way of numerous overlapping and cross-cutting communication networks. More and more people are connected in this way, giving rise to the notation of a "network society."*[50]

The advent of social media technology has had a wide-ranging impact on modern society in a number of ways. It enables those outside of news media organizations to have access to firsthand information on events, whether truly newsworthy or not, that is shared by individuals through social media channels before they are broadcasted via more traditional forums (newspaper, television, radio). Social media covers a broader global audience than traditional media with greater immediacy and very few restrictions on content. The continued engineering of improved social media tools has enabled everyone to become a publisher rather than the select few, and has allowed for freedom of personal expression at a level never before seen in human history. It has become far more difficult for governments and dictatorships to filter and restrict not only the voice of the public, but also the outside voices that they can receive.[i] While this new online freedom encourages widespread sharing of information considered relevant on any new event of interest, some call social media updates, and emerging "citizen journalism" afforded by new technology tools, an unreliable information source that lacks analysis and context.[51]

> *...people in emerging and developing nations say that the increasing use of the Internet has been a good influence in the realms of education, personal relationships and the economy. But despite all the benefits of these new technologies, on balance people are more likely to say that the Internet is a negative rather than a positive influence on morality, and they are divided about its effect on politics.*[52]

The overwhelming availability of information in modern society has enabled us to be more comparative and critical of the information we are presented from traditional sources. Individuals and groups worldwide are able to use social media tools for opinion and information sharing, and we now have a new phrase, "hashtag activism," which refers to using Twitter and other social media tools to create social awareness and movements both within local communities as well as across the globe. Vinton G. Cerf, the chief Internet evangelist for Google, states:

> *All these philosophical arguments overlook a more fundamental issue: the responsibility of technology creators themselves to support human and civil rights. The Internet has introduced an enormously accessible*

[i] Nevertheless, countries including China, Cuba, Iran, North Korea, Sudan, and Syria all prohibit access to sites such as Google, and only this year has Pakistan allowed the use of Google.

and egalitarian platform for creating, sharing and obtaining information on a global scale. As a result, we have new ways to allow people to exercise their human and civil rights.[53]

An interesting example of social networks being used for activism is the Arab Spring uprising of 2011, where people (in Tunisia, Egypt, Libya, Syria, Yemen, Bahrain, Saudi Arabia, Algeria, Kuwait, Morocco, Oman, and Jordan) protested against the governments of these countries. Although images of violent responses from ruling governments were restricted from being seen abroad, protestors were able to successfully share them with the outside world via social media. Further, social media was used to organize protests and raise awareness of the true nature of the violence against protestors.[54] Social media saw similar adoption during Iran's "green revolution" to circumvent government censorship.[55]

China's control of the outside information that the population is able to receive through the Internet is an ongoing game of cat and mouse, with the Great Firewall blocking computers and other devices in Mainland China from accessing various online services the government considers inappropriate. In July 2016, the government's Cyberspace Administration of China officially banned news sites from using social media as a source for their news reports without strict verification, stating "it is forbidden to use hearsay to create news or use conjecture and imagination to distort the facts."[56] The government also undertakes extreme measures to spread its message of propaganda, creating an estimated 488 million fake social media posts a year to distract its citizens from paying attention to negative news items and other hot topics.[57]

Russia is also using technology to control access to outside media that the state deems inappropriate. The country's *Roskomnadzor* (Federal Supervision Agency for Information Technologies and Communications) is believed to have built a massive Internet content monitoring and filtering network powered by citizens who support the country's president, Vladimir Putin. The government's level of control is such that the CEO of VKontakte, a popular Russian social media service, was forced into exile after resisting requests for data. Technology will clearly play a role in future censorship efforts, as rumors point to *Cheburashka,* the name of a system being developed to create a completely isolated network within Russia.[58]

While the use of social media for activism against oppressive regimes is a positive outcome, the engineering of an open system without careful controls also opens opportunities for the free and easy spreading of negative and harmful information and the distribution and promotion of illegal activities. The anonymity of online services like Yik Yak, a mobile app that allows for anonymous online chat within geographical boundaries, has created an open and unmediated forum for discussion.

As a result, there are many ways in which social media is now being misused, including the spread of propaganda from extremist organizations, manipulation and misguidance of underage youth, human trafficking, and the exercise of our worst human tendencies. The social media activist DeRay McKesson agrees, noting, "what social media has done is that it has exposed the intensity of hatred in America."[59] Once limited to their local community, people can now go online and join like-minded strangers in threats and bigotry against whomever they choose without fear of retribution. Hate groups are able to freely use social media technologies as effective communication, organization and recruiting tools. Rick Eaton of the Simon Wiesenthal Center notes,

> *25 years ago an organizer would have to stand on street corners handing out literature, cajole acquaintances and others to get them interested, not to mention constant phone calls to keep people interested [to] get them to rallies and so on ... Now they can easily post items to blogs and social media, send out mass emails, create discussion forums.*[60]

The growth of online hate activity has had serious consequences. Black separatist online groups have called for action against whites following recent high profile cases of black men being killed by police,[61] and white supremacy groups online have spawned a number of deadly attacks on members of black communities.[62]

In a bid to gain attention and notoriety online, many have resorted to extreme measures, such as filming illegal and atrocious acts and then posting the video and associated commentary to social media. This has emboldened people to commit crimes, reveling in the popularity harnessed online. In their own minds they have become heroes. Others have utilized social media to showcase even more heinous crimes, such as a terrorist recording video of himself murdering two police officers in France and posting the video online while holed up in the house with police outside.[63]

Recent events in Nice, France have truly laid bare the good, the bad and the ugly of social media. On the night of Bastille Day, many of those on Twitter posted heartfelt messages of support and love for those affected by the recent terrorist attack in Nice. Others filmed themselves as they panicked and ran from the scene, seemingly more concerned about shooting footage to share online than focusing on fleeing a deadly situation. Still others walked through the carnage on the ground afterwards, filming twisted bodies and sharing the video online. Twitter users then retweeted the graphic footage repeatedly throughout the night with little to no regard for the victims, their families or their friends. How can a technology that provides a venue for open and instant communication be both so positive and so negative at the same time?

There are also positive effects of the growth of online technology on society. The Pew Research Center found in a 2014 survey that 87% of users felt they were better able to learn new things, 72% reported they enjoyed an abundance of information, 75% were better informed about national news, 74% were better informed about international news, and 49% were better informed about civic and government activities in their area.[64] On the impact of social media on societal interactions it is noted that

> *...the availability of various technological platforms enables individuals to feel a deeper sense of connectivity and contribution to their social circles and the world around them. This growing dependence on social networking platforms has altered the ways in which society functions and communicates.*[65]

Social media has also been an outlet for critical information during recent US crises where the public was seeking the status of their family members or friends. During the June 2016 mass shooting in Orlando, social media was the chosen avenue to not only get the word out about the danger to those in the area but also for families to communicate with loved ones in danger. Trapped in a bathroom, some survivors recorded videos and posted them for friends and family to see, documenting the danger of the situation.[66]

An earthquake in Haiti killed an estimated 200,000 people in 2010, and immediately following the disaster, Twitter, Facebook and various blogs played a role in spreading the word and engaging millions of people in fundraising efforts. They raised $8 million for survivors within the first week. The growth of the use of Twitter was such that the television news network CNN began using it to monitor firsthand accounts coming out of the country.[67] Clearly, the Internet has become the medium of choice for most communications.

Agriculture and the Food Industry

Thanks to improvements in engineering and agricultural science, we are today, to a certain extent, able to defy Mother Nature and grow off-season crops by artificially simulating a favorable environment, without regard to the time of year and normal growing periods. Through the continued improvement of irrigation tools and techniques and engineering of other water-related technologies, we are turning otherwise dry lands and deserts into habitable and productive agricultural farmland. Although this has opened many new opportunities for food production, employment, and local economic growth, it has also left significant footprints in the surrounding ecosystems.

New and more advanced genetic modification tools and advances in chemical fertilizers, herbicides and pesticides have allowed us to grow and harvest crops at higher rates than ever before. Similarly, there are now seeds that have been engineered to resist insecticides and herbicides, add nutritional benefits, and improve crop yields. US agrochemical company Monsanto notes that in 1960 the average farmer in the US fed 25 people, while now the same farmer would feed 155 people.[68] The US Department of Agriculture (USDA) notes that wheat strains are constantly being bred and engineered to withstand diseases that might decrease their yield. Many of these genetically modified seeds are also considered to be a source of serious harm to the health of humans and the environment. In 2013, for instance, Monsanto was also found to have genetically engineered wheat, which was never approved for use, growing on a farm in Oregon. Monsanto has claimed to have dropped its development of genetically modified wheat in 2004, in response to concerns from American farmers that it would endanger exports to countries that ban such products. When this occurred, both Japan and South Korea suspended some imports of American wheat, and the European Union urged its 27 nations to increase testing. While genetically modified soybeans and corn grown in the US are chiefly for animals or used as ingredients in processed foods, wheat is consumed directly by people.[69]

Today 21% of all food calories globally are from wheat, which is providing 20% of the protein intake of 4.5 billion people in 94 developing countries.[70] Curiously, even with efficient farming, modification, and breeding of crops like wheat and soybeans, we still are unable to eradicate global hunger!

> *The world already produces more than 1½ times enough food to feed everyone on the planet. That's enough to feed 10 billion people, the population peak we expect by*

> *2050. But the people making less than $2 a day — most of whom are resource-poor farmers cultivating unviably small plots of land — can't afford to buy this food. In reality, the bulk of industrially-produced grain crops goes to biofuels and confined animal feedlots rather than food for the 1 billion hungry.*[71]

Our production of meat is on the rise, having grown from 86 million tons in 1963 to 339 million tons today. This amounts to an average meat consumption annually of almost 100 pounds per person, per year for everyone on earth. China has increased its meat consumption six-fold in 40 years, and worldwide we now feed cereals to animals that if fed to humans could sustain 3.5 billion people. With billions of farm animals worldwide, we use nearly 60% of the world's agricultural land for beef production. Not surprisingly, livestock is responsible for 18% of greenhouse gas emissions.[72] "The production of one kilogram of beef requires 15,414 liters of water on average....the production of one kilogram of vegetables, on the contrary, requires 322 liters of water."[73]

With our entire technological prowess in food production in its various forms, we are remarkably inefficient, losing or wasting one third of all food produced before it gets to the table. This is an astounding one billion tons of food.[74] In the developed world, we waste food throughout the supply chain, but particularly at the consumption stage, unlike in developing countries where most food losses occur earlier in the process. Estimates for average household food waste in Europe and North America are 95–115 kg per year, compared to 6–11 kg per year in sub-Saharan Africa and South/Southeast Asia.[75]

While a large amount of food is lost in the developing world due to harvesting problems, poor packaging and improper storage, some is simply due to a lack of refrigeration (a result of the lack of electricity). Engineers can play a role in developing simple low-cost technologies to avoid this waste. For example, Evaptainers is a startup company that has created lightweight, inexpensive mobile evaporative cooling refrigeration systems that require no electricity, and it is currently delivering them to areas in Morocco to reduce food spoilage and waste.[76]

Natural and artificial flavorings, preservatives, and other additives have not only lowered the production cost but have also allowed people to enjoy a wider selection of food with a longer shelf life.[77] Raw and produced foods are now stored and preserved as a result of new and packaging techniques – pasteurization, aseptic packaging, canning, controlled atmosphere, and vacuum packaging. While this has significantly reduced our production and storage food waste, we now suffer from waste from overproduction, and from potentially harmful health problems.[ii,78] As noted by the Natural Resources Defense Council (NRDC), 20 pounds of food per person go to waste every month, and 40% of the food produced in the US goes to waste every year.[79] According to the Environmental Protection Agency (EPA), greater than 97% of food waste generated ends up in landfills.[80] This occurs at the same time that 48.1 million Americans live in food-insecure households,[81] and one in nine people across the globe (approximately 795 million people) do not have enough food to lead a healthy lifestyle.[82]

As a result of engineering advancements in food preparation, storage, and production costs, a desire for easy, quickly prepared, tasty, and cheap meals has evolved. The fast food industry has escalated its efforts to manufacture foods and market to greater global audiences, and has shown remarkable proficiency in meeting their needs while also being sensitive to their local culinary culture and food habits. Fast food restaurants in the US now generate $191 billion annually in revenues and employ nearly 4 million people.[83] With simplified tools and automated systems for handling, preparing, cooking and packaging, the fast food industry now offers work opportunities even for the unskilled. Highly engineered point-of-sale systems now allow cashiers to simply touch the icon of the food item needed, and all traditional efforts of the cashier (calculation of the value of the total food bill) are taken care of precisely and automatically. And with the growth of technology, we will see a rapid change in the way we manufacture, distribute, and sell food. However, we will be able to bring processed food to a larger cross section of humanity worldwide.

Global Commerce

Improved transportation has contributed to a massive increase in international trade. The World Trade Organization (WTO) notes "dramatic decreases in transportation and communication costs have been the driving forces behind today's global trading system." A key component of the decreased cost of transportation is the modern shipping container, a significant engineering development that paved the way for efficient and low cost global transportation systems. From 1980 to 2011, developing economies grew from 34% of global exports to 47% and from 29% of global imports to 42%.[84] As a result of continuous improvements in global transportation systems, the components for a device can be sourced from almost anywhere in the globe, and then when

---

[ii] The FDA has an extensive list of additives and their use limits. Unfortunately, it is very difficult to regulate and many developing countries have no regulations at all.

manufactured, the device can be sold almost anywhere in the world without transportation costs overwhelming the profit margin. A community in India can have both easy access to items made on the other side of the globe, and export locally produced items to other countries - still turning a profit after shipping costs.

Economic, environmental, political, and societal concerns are no longer local problems, as the world has become better connected after moving online. Developing economies in places such as China have grown stronger, as, in order to meet high demand, production and manufacturing have been outsourced to countries with large, cheap labor pools. As a consequence, previously strong economies are beginning to lose their market share for certain types of goods.[85] The outsourcing of manufacturing has brought new opportunities (infrastructure, governance, employment) to developing countries and newly industrialized economies.

The growth of the mobile communications industry in Africa has served as a catalyst for social change by bringing the continent closer in operation to modern developed economies. Improvements include banking, (where users are able to store, accept and send money from their phone using a payment service called M-PESA), distribution of information to vulnerable rural areas, better medical service, and improved communications infrastructure. In light of the fact that Africa uses only 4% of the world's electricity, the telecom sector comprises an entirely new and significant driver of the economy, with more than 500 million phones in use across the continent.[86]

2.2 Education

The need for education has driven positive social change, as greater use of technology has increased the need for skilled workers and developing countries have seen a way out of poverty through technological growth. In a report on higher education, the World Bank notes that the need for education is strong as wages in some developing countries increase and young people strive to join the global knowledge economy. The growth in higher education in India over the past thirty years has been remarkable. The World Bank reports that general education colleges have grown in number from 3,421 in 1980 to 10,377 in 2005, while universities deemed "of national importance" have grown from 110 in 1980 to 364 in 2005.[87]

Between 2000 and 2010, African higher education enrollment more than doubled, increasing from 2.3 million to 5.2 million. There were 200 public universities and 468 private higher education institutions in Africa as of 2009. Overcrowding is common, with 50% more students per instructor than the global average.[88] Similar growth has been observed in China, with the number of private higher education institutions rising from 39 in 1999 to 727 in 2014. In 2015 it was reported that there were 2,845 higher education institutions, even after the combination of hundreds of institutions in the 1990s.[89] In China, nine million students each year take the 'gaokao,' or university entrance exam, with admissions slots to universities available only to roughly two-thirds of those students. Technical education is considered culturally to be very important there also, as China seeks to rise on the world stage as a technological and economic superpower.

A broadly educated workforce for economic growth demands that we educate our girls at the same rate that we do our boys. The United Nations Girls' Education Initiative (UNGEI) notes that in global primary school enrollment, for every 100 boys out of school there are 122 girls out of school. In developing countries the statistics can be very different. In India 426 girls per 100 boys are out, and in Yemen 270 girls for every 100 boys.[90] As societies change over time and education increasingly goes online, gender imbalances in communities around the developing world may begin to shift. The United Nations (UN) notes that the growth of technology in education should have a transformative effect, bringing education to women in developing countries, and closing the educational gap between men and women.

Students and educators are better equipped today for teaching and learning due to technological advances in personal computers, the Internet, social media, mobile phones and tablets, and related educational technology resources. This has had a significant impact on teaching and learning methodologies in our education system, with easier collaboration, information gathering, teacher preparation, and remote learning.[91]

Traditionally, our education system (particularly K-12 education) was based mostly on textbook learning. Textbooks in K-12 education today mainly serve as an introduction to the subject matter. In addition to the textbook, which in some cases is electronically enhanced, many educators and students rely on the Internet to supplement instruction. In short, the teacher used to be the primary source of information in the classroom, and now the spotlight is shared between the instructor and a source of worldwide shared knowledge (the Internet).[92] The instructor now serves as a moderator, introducing the students to the subject and guiding them on where and how they can find information to educate themselves further. Educators are able to be more broadly prepared with a variety of background information before entering a classroom, so they can relate their teaching to most relevant current affairs, giving students a better perspective on the relevance and importance of the chosen topic.

The downside of technology adoption in the classroom is the impact on student-teacher interactions. It can also have a negative impact on the students' communication ability and creativity, since solutions to common assignments are often readily available online.[93] This raises the question of whether we want our future generations to be educated by largely unregulated sources on the Internet and social media. There is currently little to no control over the quality and veracity of what is being shared, and the underlying motivation behind the person or organization putting forth the information may not always be clear.[94]

In light of the overwhelming access to information online, some educators are today emphasizing pedagogy more than content. For example, through various forums for learning (such as open educational resources (OER), free online courseware, TED talks, podcasts, YouTube, and open bibliographic databases, individuals can access some of the world's greatest thinkers directly online without the requirement of enrollment at a university or institution of higher education. The only requirement is a device and a connection to the Internet. Does this mean that in the future, the demand for highly knowledgeable educators will diminish and instead educational institutions will call for tech-savvy individuals with great pedagogical skills?

Furthermore, new technology tools are continuously being designed to facilitate learning and best meet the needs of the individual, regardless of physical or learning challenges. Assistive technologies include tools to improve mobility, hearing and visual aids, computer-based instruction, physical positioning, optical character recognition, word-prediction programs, speech synthesis tools and more.[95]

As the rise of technology and communications has led to networks capable of transmitting high quality video and large data files, our society has begun to utilize online learning and remote training programs commonly referred to as eLearning. The use of this new educational tool has resulted in a number of positive effects, including:[96]

- Students are more engaged and able to develop 21st century skills.
- Teachers have a more positive attitude toward their work and are able to provide more personalized learning.
- Family interaction and parental involvement may increase.
- Communities benefit from bridging the digital divide. Economically disadvantaged students and children with disabilities benefit particularly.

- Economic progress can result from direct job creation in the technology industry as well as from developing a better-educated workforce.

While most attitudes are positive, social concerns have also been raised about eLearning. While the assumption is that all young people embrace and understand technology completely, the "digital divide" between students may mean that their learning experience is challenged by the integration of technology, by virtue of their lack of technology skill, or by lack of access to appropriate technology for learning purposes. An overdependence on technology may occur where instructors eschew face-to-face meetings and replace them with prerecorded material online. Interactions then occur only through discussion and messaging tools within learning management systems. Poor course design and setup can also cause problems for learners online, particularly as they may not be relying on textbooks with clearly demarcated learning segments.[97] Finally, some believe eLearning may result in "dehumanizing" education.[98]

Nevertheless, one version of e-Learning, Massive Online Open Courses (MOOCs), has scaled free education across the globe. After early successes in the US, the idea has spread around the world, with the following just a short listing of global MOOC providers offering free and low cost education to their citizenry and beyond.[99] It is of note however that most of these offerings are from and for developed countries – for instance, there is not an African MOOC on the list. One of the initial organizations offering free online courses in the US, Coursera, made a bold promise in 2012 to provide free education from top universities to empower people with an Internet connection from across the globe. As of 2016, Coursera now offers 1,563 courses from 140 partners across 28 countries, with many of their courses still accessible for free.[100] A far from comprehensive list of global MOOC providers includes:

- Coursera (US)
- EdX (US)
- Udacity (US)
- Khan Academy (US)
- OpenUpEd (Europe)
- Iversity (Germany)
- FutureLearn (UK)
- Swayam (India, in launch phase)
- ClassCentral (Hong Kong)
- Chinese MOOCs (China)

- Edraak (Arab world)
- FUN: France Université Numérique (France)
- IndonesiaX (Indonesia)
- Malaysia MOOCs (Malaysia)
- MiríadaX (Spain)
- Open2Study (Australia)
- Open Education (Russia)

While much debate is still underway as to the MOOCs' effect on education in the developed world, a study recently found that they are having an outsized effect in developing countries. 49% of MOOC users polled in Colombia, South Africa and the Philippines stated that they had received certification for at least one course, and nearly 80% of all MOOC users in these three countries said they had completed at least one course. By contrast, US and European individual course completion rates are typically only in the 5-10% range. The study's lead researcher, Maria Garrido, notes that:

> *Many people assumed that in developing countries, MOOCs would only be used by the rich and well-educated ... We were excited to find that this is not the case. Many users come from low- and middle-income backgrounds with varying levels of education and technology skills.*[101]

Student Mobility

As a result of improved global mobility, over time, greater numbers of young people have relocated to new countries to gain a better education than they may be able to in their own countries. This has created global competition for admission to educational institutions on foreign soil throughout the world.

The US has the world's largest international student population, reported to be 974,926 students as of 2015. China and India send the most students to the US, with a total of 436,928 students for the two countries in 2015. While the percentage of Indian students coming to the US has been steady since mid-2000 (13.5% in 2006, 13.6% in 2015), their numbers have almost doubled. There was a significant increase in the share of Chinese students since mid-2000 (11.1% in 2006, 31.2% in 2015), and their numbers have increased fivefold.[102] Unsurprisingly, both of these cultures put a very strong emphasis on educational attainment for young people, and both have difficult and lengthy university entrance exams, making the US an attractive location for students from these countries.

Similarly, the number of American students going abroad to study has increased over the last 10 years. While 223,534 studied abroad in 2006, the number rose to 304,467 in 2015.[103] Many international students coming to the US are motivated by the opportunity to break out of their socioeconomic status back home and are driven by culture and family to take advantage of all available opportunities for educational success and gainful employment. In contrast, students in the US and Europe typically cross global borders in search of international education to gain experience, see the world, and experience foreign culture.

According to a study by the Brookings Institute, two-thirds of international students coming to the US are enrolled in Science, Technology, Engineering and Mathematics (STEM) or business-related fields, compared to half of American students.[104] As noted in *S&E Revitalization, A New Look*, cultural and familial forces are strong for these young people. For those from China, India and other emerging economies, engineering in particular is a respected and popular subject for study, considered to guarantee success in return for hard work and study.[105] A vast majority of international students in the US come from large cities in emerging economies, and their interest in STEM education may be a side effect of observing the growth of new markets rooted in engineering and technology within their native countries.

Within the US, as in many other developed countries, we now have greater social awareness of our pressing global challenges by virtue of our improved communications technologies, whether they are economic, technological, social, or environmental. We are increasingly reliant on advanced technologies, and as a result, there is an increasing demand for technically skilled people to build, maintain and use new technologies across almost every industry. While there is much discussion on this point, advocates state that in the US the demand for STEM professionals will increase significantly over the next four years, and they predict there will be a million extra STEM jobs in the workforce by 2020. They also note that STEM jobs pay an average of 10% higher than non-STEM jobs with the same educational requirements.[106]

With an overabundance of programs pushing the idea of increased student enrollment in STEM education in the US, we are actually capable of adding many more STEM workers to the workforce. With much of our production and associated engineering work still being outsourced to emerging economies, there could be a significant challenge in the future in finding a balance between the number of qualified engineers and opportunities in our domestic job market.

2.3 Energy

> *While the overall energy history of the United States is one of significant change as new forms of energy were developed, the three major fossil fuels - petroleum, natural gas, and coal, which together provided 87% of total US primary energy over the past decade - have dominated the US fuel mix for well over 100 years.*[107]

Coal has played a prominent role in our history and continues to be utilized, despite its contributions to climate change as a major contributor to CO_2 pollution in the atmosphere. It formed the foundation of the Industrial Revolution, providing power for steam engines, furnaces and other machinery and becoming a part of everyday life. Coal lamps, coal furnaces, and coal locomotives all used this cheap and efficient source of power, despite its obvious failings in the form of particulate and CO_2 pollution. Coal dust was a common killer, and due to poor working conditions coal workers were involved in labor strikes more often than in any other industry. Coal is still used today for the generation of electricity, polluting the air in poorer communities while supplying 50% of the electricity used in the US.[108]

But as coal continues to be mined and used as fuel across the globe, what price are we paying in terms of human health? Researchers from the University of Sydney found that those in close proximity to coal mines suffer from lung cancer, chronic heart and respiratory diseases, kidney problems, birth defects, and high levels of heavy metals.[109] A study in South Africa found that acidic coal mine drainage caused problems with water sources, acid rain is formed from coal air pollution, and coal mining discard-heap fires release toxic compounds into the air.[110] Coal-fired power plants in India, a country where half of all electricity generated is from coal, are believed to be causing more than 100,000 premature deaths every year from fine particle pollution, with 10,000 of those victims being children under 5.[111]

Developing countries typically have a very different mix of energy sources than developed countries, often at a significant human cost. Kerosene is a popular fuel, yet is quite expensive, providing light and heat but harming human health due to black carbon pollution. Coal is another source of cooking fuel, but as noted it too has a multitude of associated health problems. Biomass and animal waste are also used widely, even though the combustion of dried animal dung results in fine airborne particles that have been found to cause acute respiratory infections, chronic obstructive lung disease, and cancer.[112] Four million people die prematurely each year from illnesses relating to use of biomass and coal as cooking fuels.[113] In some particularly poor areas of the developing world, people burn trash, including plastic, as a cooking and heating fuel, the health dangers of which are extraordinary.

> *In developing countries, especially in rural areas, 2.5 billion people rely on biomass, such as fuelwood, charcoal, agricultural waste and animal dung, to meet their energy needs for cooking. In many countries, these resources account for over 90% of household energy consumption....About 1.3 million people - mostly women and children - die prematurely every year because of exposure to indoor air pollution from biomass.*[114]

In developing countries, the widespread implementation of renewable energy is crucial. Enormous investments in building renewable energy sources are required to have any hope to fulfill the promises made at the COP21 summit. This would typically be in the form of solar energy, due to its relative affordability and ability to be implemented in smaller installations, even down to the individual home. But there are problems with this form of renewable energy. Solar panels are expensive and inefficient (below 25% efficiency). Also, power from solar farms must be accommodated on existing transmission lines. Preexisting electrical wiring in some developing countries is already a patchwork mess of stolen electricity and jerry-rigged cables. The Prime Minister of India's goal, for example, is that renewable energy will provide 175 gigawatts of power by 2022 – an extraordinary goal requiring investment of around $90 billion.[115]

It is obvious that energy demands throughout our modern world are only going to grow. With the widespread adoption of technology our global energy consumption has gone from an estimated 54,335 terawatt-hours in 1973 to 104,426 terawatt-hours in 2012. It has been estimated that from 1998 to 2008 the average energy use per person globally increased 10% while the world's population simultaneously grew by 27%, adding many more energy consumers. During the same time period, energy use in the Middle East increased by 170%, China by 146%, India by 91%, Africa by 70%, Latin America by 66%, the US by 20%, and the EU-27 block by 7%, while the world's overall usage grew by 39%.[116]

So far renewable energy technologies represent only a tiny fraction of all energy generated worldwide. Coal/eat represents 41.3%, natural gas is 21.7%, hydroelectric is 16.3%, nuclear is 10.6%, oil is 4.4%, and others, of which renewables form a part, are just 5.7%.[117] Without massive technology improvements in renewable energy technologies and drastically lowered costs, fossil fuel utilization and its associated social impact is here to stay.

Growing Demands

Energy demands of the manufacturing sector are growing, where machines are increasingly enhancing or displacing human labor. In the food industry, energy needs are increasing for all steps of the process - growth, processing, packaging, distribution, storing, preparing, serving, and disposing of food.[118] The National Academy of Sciences (NAS) reports that industrial energy needs in the US are projected to grow by 22% during the next 25 years, primarily in petroleum refining, chemicals, and the paper and metal industries.[119] Similarly, energy demands are also growing in the transportation sector as goods are shipped globally from manufacturing centers around the world. In the US, we are more mobile today than ever before, with 28% of the energy in the country being used in the transportation sector for people and goods.[120]

As incomes rise across the developing world, the amenities of a modern existence move within reach. While many focus on greater use of lights, computers, televisions, etc., the greatest future energy requirement in developing countries is predicted to be from air conditioning systems. Compared with the United States, India's population is four times as large, representing a significant challenge in itself. With its warmer climate, the country also experiences an average of more than three times as many cooling degree days (CDDs) per person, a measurement of the number of days each year where temperatures are above 65 degrees and air conditioning is typically used.[121]

While just 2% of Indian homes currently have air conditioning, compared to 87% in the US, air conditioner sales growth has been reported at 20% per year. One estimate puts the possible increased energy demands of air conditioning just in Mumbai at one-quarter of the current energy used for air conditioners for the entire US.[122] Air conditioning in Hong Kong has been projected to account for some 60% of the electricity usage in the summer months.[123] Air conditioner efficiency will be a key research area as the number of units is projected to increase from 900 million to 3.5 billion by 2050.[124] China is also experiencing significant growth, with sales of air conditioners almost doubling over the last five years. There were 64 million units sold in the country in 2013, more than eight times the number sold in the US.[125] The increased demand for electricity to provide air conditioning for large populations in developing countries will seriously challenge efforts to reduce fossil fuel usage and associated emissions.

The use of electric vehicles is so far having only a minor impact in the developed world, with high cost and a lack of infrastructure a deterrent to mass adoption. However, companies such as Nissan, and organizations such as NIST, have been exploring the use of electric vehicles to re-charge the grid. Nissan launched a pilot program in the UK where 100 of its Leaf electric cars were integrated into the power network. In such a system, drivers are able to charge their cars as normal during off-peak times, but the charged batteries would provide added capacity to the national grid, helping balance demand at peak times.[126]

Fossil fuels have played an integral part in enabling our existence on this planet, as well as powering technological advances via increased manufacturing and production. However, they have also increased our global carbon footprint and negatively affected climate change. We now face a future where it is imperative to reduce fossil fuel consumption and to engineer improvements such that they burn cleaner and more efficiently. We must also supplement them with energy created by greater use of renewable energy technologies, as we must drastically reduce our carbon emissions.

2.4 Culture

> *In order to bring about deep economic and social changes and promotion of the living standard as well as filling the gap between themselves and the developed countries, the developing countries are in need of science and technology, and development has become an important factor for industrial and economic progress. But science and technology have not been created and developed in isolation and introduction of any new technology is a cultural phenomenon, directly affecting the cultural values and the behavior of communities.*[127]

Culture typically describes social norms based on lifestyle, values, ideologies, and economic status of a group of people within a community and/or geographical region. Social change raises not only questions but also fear and concern. Humans have the intrinsic ability to adapt to change over time; however, our ability to accept and adapt to new ideas is often limited by our surrounding social and cultural environment.

While technology influences culture and human behavioral pattern, culture serves both as a limiting and contributing factor in adoption of new technologies and modern amenities. Thus, it is important to understand culture's role as a catalyst in contributing to technology-driven social change.

Unfortunately, the role of culture is often neglected when considering adoption of new technologies that in the big picture contribute to social change. A society's readiness to adopt new technologies often depends on its impact in terms of safety, welfare, economy, and social values and ideologies. One example is autonomous vehicles. The notion of a self-driving automobile may sound very exciting and futuristic and may seem to offer tremendous convenience. However, the idea of autonomous automobiles driving on public streets raises many concerns related to safety, liability, system security, etc.[128] While running a pilot program with a handful of autonomous vehicles on the local streets may be easily accepted in a place like San Jose/Silicon Valley, California, where the culture has been shaped by innovators and entrepreneurs, such a program may experience great pushback in El Paso, Texas, where bleeding-edge technology is not an integral part of the local culture.[129]

There are other ways culture affects the adoption of technology. For instance, in a factory the culture among the workers plays an important role in their readiness to adopt new tools and technologies. Automation of manufacturing plants on one hand promises to increase the overall output of a factory and generate increased revenue, but at the same time takes away jobs from workers. In industries where automation has reduced employment, such as manufacturing and assembly, workers may have a much more negative outlook on the adoption of new technology than those working in a biotechnology firm, where new technology may increase the number of employees required. "We often accept an innovation owing to its evident utility at the individual level and then criticize it for its consequences at the collective or cultural level."[130]

A Tale of Two Traditional Cultures

Not only do our traditions and the way we communicate differ from culture to culture, but so do our adaptability to new technologies and the level of change we are willing to accept. The culture of the Amish people of the US and Japanese people are two good examples of this, and provide for a striking contrast. Both are rooted in long-standing traditions that have been carried down over many generations and that define the individual. However, the contrast between these two cultures in their adoption of modern thinking and new technologies is significant.

Japan today is considered to be one of the world's most technology-forward countries, wherein technology has left significant footprints on its economy, infrastructure, and lifestyle. Yet, the Japanese people still hold on to traditions that are deeply embedded in their culture. For example, Japanese people traditionally bow to each other before shaking hands, do not wear shoes inside Japanese homes, and hand over business cards with two hands. These are just a few examples among many of traditional behavior, but the process of keeping to traditions and of maintaining cultural cohesiveness has not stopped the Japanese from adopting modern technologies. Instead, they have utilized modern technology at its best to help them maintain traditions. For instance, cleanliness is an important part of the Japanese culture.[131] A good example of how the Japanese have used technology to maintain this tradition and lifestyle is the way they design their bathrooms and toilets to ensure maximum cleanliness, utilizing a combination of multiple devices and technologies. The Japanese toilet is a marvel of high technology and enhanced functionality in comparison to a Western toilet. Japanese culture celebrates the toilet to such an extent that Toto, a major manufacturer, has recently opened a $60 million museum devoted purely to its toilet products.[132]

> *The Japanese are just as fastidious at home as at play. Cleanliness matters. Most have at least one bath a day; rare is the young woman who does not have at least two. Washing does not involve a superficial flick of the flannel, but a vigorous all-over scrub, often with an extremely rough nylon towel. And this before a person gets into the bath.*[133]

The Japanese encourage the use of robots in their culture, with applications including elder care, disaster relief, portrait drawing, and a variety of other uses. Robots that many in the West would see as toys, like Sony's classic Aibo robotic dog, or many of the hundreds of new humanoid robots being sold today, are treated as if they are living pets. In Japan, robots are considered to have a spirit or soul. The concept is consistent with the country's primary religion, Shinto, which supports the idea of animism, where both animate and inanimate objects possess a spirit or soul.

> *Given that Japanese culture predisposes its members to look at robots as helpmates and equals imbued with something akin to the Western conception of a soul, while Americans view robots as dangerous and willful constructs who will eventually bring about the death of their makers, it should hardly surprise us that one nation favors their use in war while the other imagines*

them as benevolent companions suitable for assisting a rapidly aging and increasingly dependent population.[134]

In contrast, the culture of the Amish people in the United States shows how culture can serve as an anti-catalyst for the adoption of modern technologies. The lifestyle of the Amish people and the values they practice and preach are dictated by a number of rules known as the Ordnung, which were outlined before electricity was discovered.[135] In many ways the lifestyle of the "modern" Amish is reminiscent of the early settlers in the Midwest of the US in the 1800s, often with a very traditional low technology existence. The Amish culture strictly opposes certain modern technologies and instead emphasizes traditional processes and values, stressing their importance in maintaining their cultural identity. The Young Center for Anabaptist and Pietist Studies at Elizabethtown College in Pennsylvania clarifies a widely held belief that the Amish eschew all technology. Rather, "the Amish do not consider technology evil in itself but they believe that technology, if left untamed, will undermine worthy traditions and accelerate assimilation into the surrounding society."[136] Consequently, they limit mobility by keeping to horse and buggy for transportation and limit exposure to outside society by not allowing television and the Internet. They do, however, use electricity and selected powered appliances and tools, but use their own sources for electricity generation (solar, etc.) rather than outside electricity service lines.

The comparison of Japanese and Amish cultures describes what Danila Bertasio puts forth in her examination of Western and Eastern traditions and their connection to technology as the two souls of culture: one which is closely bound to the humanistic tradition and one which is open to scientific-technological innovations.[137]

India: Technology and Culture

India is another country where culture has played an important role in the implementation of technology. India is today considered a "newly industrialized" country with a growing economy following growth of the country's GDP and increased foreign investments.[138] The total population of India is more than 1.2 billion, with nearly 70% living in villages.[139] The urbanization India has experienced in the last two decades is reminiscent in many ways of the urbanization that took place across Europe in the 19th century.

Today, city life in prosperous parts of India is in many ways similar to that of cities in the United States or across Europe. Ambitious young people are working hard and thriving in their white-collar careers. Fast food, mobile communications, and social media are an important part of their existence in a modern, global, always-connected city. While tradition and culture are still valued and highly regarded, people in metropolitan cities within India are very open to the new ideas, technologies, and amenities that constitute a "modern" existence. Thus, they try to maintain a balance between accepting new technologies and staying connected with their cultural heritage.

A desire to preserve culture and tradition has in many ways served as a driving force among Indian engineers and entrepreneurs. They are delivering new technologies that best meet the cultural need of the country's growing number of middle-class consumers. An example is the assimilation of technology into daily routines. Professionals running short on time in the morning hours can now download an app on their smartphone that allows them to perform their daily prayer, in lieu of physically attending a local temple.[140]

In rural India, where tradition and culture are often considered all important, adoption of technology has not been as swift and far-reaching as in the cities. A common epithet is, "why change something that has worked just fine for generations?" In these communities, using a mobile app to pay your respects to God instead of going to the temple is out of the question. One is likely to hear that "God lives inside the temple and not inside your iPhone!"

Further, in contrast to metropolitan cities where mobile phones are widely used among young people (particularly among young girls, where they are considered as a necessity to help keep them safe), it was recently reported that in Bihar women have been banned from using mobile phones. It was believed that they "debase the social atmosphere" by allowing women to organize the process of eloping.[141] According to a village head in Gujarat, where unmarried women are banned from using mobile phones, "Why do girls need [a] cell phone? Internet is a waste of time and money for a middle-class community like ours. Girls should better utilize their time for study and other works."[142]

These views are shared by many in rural India, where local culture typically dictates that males exercise authority and decree what is best for their community. A culture's infrastructure, whether emphasizing hierarchy or individual governance, has a great impact on its adaptability to change. While the local culture in rural India may be more hierarchically structured, the culture in the cities focuses more on the well-being of individuals and their own progress, with less emphasis on a traditional communal thought process.

The Right to Bear Arms

An important example of culture pushing back on engineering is the volatile situation surrounding gun ownership in the US. While many countries in the rest of the world advocate for banning the possession of firearms, the Second Amendment of the Bill of Rights states that citizens have the right to "keep and bear arms."[143] Carrying a gun is not only considered to be a civic right, but has also become a part of the local culture in many parts of the United States. In certain states, it is also allowed by law. At the 2016 Republican National Convention in Cleveland, Ohio, regulations did not allow tennis balls, metal-tipped umbrellas, or canned goods in the event zone, but guns were allowed under Ohio's open-carry laws.[144]

At 112.6 guns per 100 residents, US gun ownership rates are the highest globally, almost double the rate of Cyprus in the #2 position.[145] Of the 650 million guns owned worldwide, 41% are owned by Americans.[146] Unsurprisingly, we are also the largest exporter, representing 31% of all global arms exports in 2014.[147]

In the United States in 2013, there were 33,169 deaths due to firearm use and misuse, coupled with 84,258 nonfatal injuries from guns.[148] The June 12, 2016, killing of 49 people in Orlando, Florida, with a high-powered assault rifle, sparked a renewed debate over the need for guns and the lack of control over their purchase and usage. Following the attack, an effort to create policy that restricted purchasing of weapons such as those used in the shooting was unsuccessful. Four different bills proposed in the United States Senate to limit gun sales were rejected in June 2016.

Engineering has a clear role in preventing many of the yearly injuries and deaths from firearms, but the gun industry, pro-gun organizations such as the National Rifle Association (NRA), and individual gun owners resist the use of technology to control use. Although smart guns with technology-based trigger lock mechanisms have been engineered and shown for more than a decade, they have not gained traction with consumers, as the culture pushes back on the technology. American gun owners and purchasers believe their constitutional rights are being threatened by the development of smart gun technology, as well as legislation to require it on new firearms.

> *The Justice Department gave Colt a $500,000 grant in 1997 under the Clinton Administration to complete development of a handgun that would work with RFID (radio frequency identification) via a wristband. Smith and Wesson received more than $3 million in DOJ*

grants between 2000 and 2004 to develop smart guns for law enforcement. The prototypes were completed but no one was using them and Smith & Wesson had to layoff 15 percent of its staff due to boycotts.[149]

Colt and Smith and Wesson suffered revenue-crushing boycotts after developing their government-sponsored smart gun prototypes. In the years since, gun dealers who have expressed interest in selling smart guns alongside traditional firearms have had to cancel their orders due to boycotts and fears of bankruptcy.[150]

So, the question becomes this: How can the US protect its citizens against gun-related deaths and injuries when the solutions to many of these preventable deaths have already been engineered and are readily available? What is the solution when the culture rejects them, even in the face of regular and devastating gun violence throughout the country? For now, there is no answer.

2.5 Why Is Philanthropy Important to Engineers?

Continuous globalization and improved communication have connected people and their communities from across the globe, broadening the average engineer's outlook and sparking an interest in contributing to the welfare of others through use of engineering skills. This increase in social awareness and responsibility engages the modern engineer in activities outside the scope of traditional engineering. Young engineers often find it more fulfilling to be making a positive social impact, instead of just working towards increasing profits for their employer.

Representing the majority of organizations engaged in creating positive social change, the nonprofit world is a significant part of the US economy. There are an estimated 1.4 million operating nonprofits in the US, employing approximately 11.4 million workers, or 10.3% of the nation's workforce.[151] Nonprofits represent 5.4% of GDP, contributing $905 billion to the US economy, and they expended $2.1 trillion in 2013, with an estimated $5.1 trillion in assets.[152] Consequently, engineers have a significant role to play in this environment. While scholars in social science and policy makers play an important role in identifying and understanding our world's most pressing social and economic concerns, as well as educating society on how we might use policy to address these concerns, designing sustainable solutions to address these problems often requires significant technical expertise.

Engineers Engaging in Philanthropy

There are a number of organizations that are engaging in engineering work to improve society. The first of these, and perhaps the most recognizable, is Engineers Without Borders (EWB). They describe their mission as building "a better world through engineering projects that empower communities to meet their basic human needs and equip leaders to solve the world's most pressing challenges." [153] With approximately 300 chapters in the US and an estimated 17,000 total student and professional volunteers, the organization has been involved in many projects with social impact throughout the world. Sustainability, accountability, ownership, leadership, education, and community partnerships are all stated as important components of the organization.

> *Our approach to development is based on more than blueprints and measurements; it's based on real relationships and five-year partnerships with communities. We do more than build latrines for communities–we equip them to build and maintain latrines themselves. At the same time, the volunteers and community members learn valuable leadership skills by charting pathways through complex challenges and achieving shared goals.*[154]

They also offer the services of their most skilled volunteers via the Engineering Service Corps to organizations involved in international development activities. Representing another large collection of engineers with a development engineering and social change mindset, Engineering For Change (E4C) is an online community of 20,000+ members founded by the Institute for Electrical and Electronics Engineers (IEEE), Engineers Without Borders (EWB), and the American Society of Mechanical Engineers (ASME). The organization provides an online academy, an innovation laboratory and a media platform that demonstrates the engineering challenges and solutions found in the developing world, sharing methodologies and approaches to a large group of motivated engineers and those connected with engineering throughout the world.[155]

Government Organizations

There are multiple government organizations working in the US and abroad that provide funding for and engage directly in development engineering projects and philanthropic work globally. One of the most visible is the United States Agency for International Development (USAID). In 2015 they provided $22 billion to partner organizations around the world to "end extreme global poverty and enable resilient, democratic societies to realize their potential."[156] The agency states its goals are to:

- Promote broadly shared economic prosperity,
- Strengthen democracy and good governance,
- Protect human rights,
- Improve global health,
- Advance food security and agriculture,
- Improve environmental sustainability,
- Further education,
- Help societies prevent and recover from conflicts, and
- Provide humanitarian assistance in the wake of natural and man-made disasters.

As part of their Higher Education Solutions Network project announced in 2012, USAID provided $130 million in funding to establish global research and development centers in development work at seven top American and foreign universities. These included the Massachusetts Institute of Technology (MIT), the University of California – Berkeley, Michigan State University, Duke University, Texas A&M University, the College of William & Mary, and Makerere University in Uganda.[157] There is a significant engineering component to this work: geospatial mapping, large-scale technology and codesign, drought solutions, development technology evaluation, technology innovation networking and food supply solutions, among others.

Another government organization involved in development engineering is The Global Environment Facility (GEF), an interagency group established on the eve of the 1992 Rio Earth Summit for the purpose of tackling major environmental challenges worldwide. The organization has provided $14 billion in grants over the past 24 years, and secured $75 billion in additional financing for almost 4,000 projects. The GEF has grown to become an international partnership of 183 countries, international institutions, civil society organizations, and the private sector to address global environmental issues.[158] Examples of current GEF projects include the development of energy efficient lighting, the improvement of energy resilience, use of biomethane, biomass-based electricity generation, green chemistry and associated technologies, e-waste treatment, sustainable urban systems management, mini and micro hydropower plants, and treatment of mercury pollution in mining wastewaters.[159]

While the United Nations strives "to achieve international co-operation in solving international problems of an economic, social, cultural, or humanitarian character," specific organizations within the larger structure of the UN work directly on these issues. The United Nations Development Programme (UNDP) works in 170 countries on issues of poverty, inequality, and exclusion, joining with countries to achieve sustainable development through improved policy, partnering, and growth of institutional capabilities.[160] The budget for UNDP projects in 2016 was $4.9 billion, with 3,985 active projects around the globe funded from a variety of sources, including the GEF, Japan, Argentina, Europe, Germany, the UK and many other countries. Current UNDP projects include work in the areas of responsive institutions, inclusive and sustainable growth, democratic governance, crisis prevention and recovery, climate change and disaster resilience, development impact and effectiveness, and gender inequality.[161]

Corporate Philanthropy

Corporate philanthropy is defined as "investments and activities a company voluntarily undertakes to responsibly manage and account for its impact on society."[162] This may be in the form of money, products, services, or employee volunteering, usually to advance a social cause either directly or through the work of a non-profit organization. Corporations benefit from engaging in philanthropy, as it increases the value of corporate stock, improves company public relations, establishes strategic partnerships, and more.[163] The growth in popularity of corporate philanthropy has given rise to corporate foundations, most notably through the many visible successful technology companies who see a social benefit from their shareholders, other stakeholders, employees and customers. We consider a few examples here.

The Intel Foundation granted $39 million in FY2014. It has supported the Intel Science Talent Search and the Intel International Science and Engineering Fair, and has also provided STEM grants in communities where the company has a presence. The company achieved one million hours of employee volunteering in 2015, with $10 being donated to schools and community organizations for every volunteer hour.[164]

The GE Foundation granted $108 million in FY2014 in the areas of health, education and disaster relief. Their Safe Surgery 2020 initiative aims to improve safe surgery access through training, infrastructure, and innovation to the five billion people worldwide who lack access. The

Foundation also works on K-12 education grants and workforce development. Their Developing Health program is also improving healthcare access for some of the world's most vulnerable groups in 16 countries.[165]

The Exxon Mobil Foundation granted $75 million in FY2014 across three key focus areas: malaria, math and science, and economic opportunities for women. The Foundation supports prevention, treatment, and research and advocacy programs with the goal of eradicating malaria from the planet. Math and Science programs are also an important component of Exxon Mobil's philanthropic efforts, with their engineering-focused programs supporting the development of STEM programs and working to attract more female and Hispanic students to engineering. The Foundation's Women's Economic Opportunity Initiative has helped tens of thousands of women in 90 countries by providing training programs, access to technology, and networking opportunities.[166]

Private Philanthropy

Private foundations in the US have grown over recent years, and the Bill and Melinda Gates Foundation is the largest example, with an endowment of $44 billion. These organizations can wield tremendous power, and they employ thousands of people in attempting to address society's biggest issues. Engineers work closely with policy makers and social entrepreneurs in private foundations, addressing pressing concerns related to sustainable development in health, water and energy.[167] The Foundation is addressing issues with partner organizations in four key areas:[168]

- Global Development Division works to help the world's poorest people lift themselves out of hunger and poverty.
- Global Health Division aims to harness advances in science and technology to save lives in developing countries.
- United States Division works to improve US high school and postsecondary education and supports vulnerable children and families in Washington State.
- Global Policy & Advocacy Division seeks to build strategic relationships and promote policies that will help advance the Foundation's work.

A typical engineering challenge undertaken for the developing world by the Bill and Melinda Gates Foundation is seen in their collaboration with the University of Toronto in 2011:

> *Design a toilet that's off-the-grid – no running water, no sewerage system, no electricity. Make sure it's self-contained: human waste goes in; clean water, carbon dioxide, mineral ash (for fertilizer) and energy comes out, in about 24 hours. Oh yes, and it has to work for only five cents per user, per day.*[169]

The Foundation hosts a group of initiatives known as the Global Grand Challenges, seeking to provide grants for innovative solutions to a wide variety of global health and development issues. Examples of current challenges include improved health analytics, extreme low cost point-of-care testing, low cost vaccine manufacturing and delivery, malaria diagnostics, microbiome engineering, digital financial services data analysis, malnutrition status assessment, and simple and effective mobile money market solutions. Bill and Melinda Gates note in their 2016 open letter that the major challenge the planet faces is in the area of energy, both in the adoption of newer technologies to stall or reverse the effects of climate change that energy production has wrought and to bring energy to the billions who currently do not have it.[170]

We are living in a world transformed from a hundred years ago, and technology is now a deeply ingrained part of most cultures. Along with education, cultural forces and energy needs, these areas have acted as catalysts for social change where engineering has played a significant role. Social media has changed the way we take in and disseminate information, increasing the speed and expanding the reach of our communications. Citizen journalism has become a pervasive new reality, as we are now able to widely share rich media immediately without the normal societal filters of morality or good taste. Technology is continuously improving the world's access to education, decreasing the cost and moving it online, so that students in both the developing and developed worlds can build their knowledge outside of their traditional school systems. Our food systems have also been improved with technology, enabling greater amounts of food production, transport and storage than ever before, even though our ability to feed those in need throughout the world is still lacking. New demands for energy are at odds with our efforts to combat climate change, so engineers have been working to create cheaper clean energy solutions. As these technologies have been developed, culture has been instrumental in determining their success. The growth of philanthropy, particularly in US culture, has resulted in large organizations that are expending millions of dollars on creating positive social change through the judicious application of various technologies. When developing the technologies of the future, engineers must be careful to appreciate the importance of culture and engage non-traditional approaches, such as leveraging philanthropic resources to affect positive social change.

References

[41] Creatista, "Group of 8 teenage girls text messaging at an amusement park", Digital Image, Shutterstock, 2016, accessed at http://www.shutterstock.com/pic-115122034.html

[42] Fairphoto, "Entering Agbogbloshie", Digital Image, Flickr, February 28, 2014, accessed at https://www.flickr.com/photos/fairphone/12830951233

[43] Author's View.

[44] "6th Annual Sustainable Innovation Forum", United Nations Environment Programme, 2015, accessed at http://www.cop21paris.org/

[45] McKeever, B., "The Non-Profit Sector in Brief 2015: Public Charities, Giving and Volunteering", The Urban Institute, October 29, 2015, accessed at http://www.urban.org/research/publication/nonprofit-sector-brief-2015-public-charities-giving-and-volunteering

[46] "Calling a country far, far away", The Telecommunications History Group, Inc., accessed at http://www.telcomhistory.org/vm/scienceLongDistance.shtml

[47] Winsor, M., "Mobile Phones in Africa: Subscriber Growth to Slow Sharply as Companies Struggle to Reach Rural Population and Offer Faster, Cheaper services", *International Business Times*, October 20, 2015, accessed at http://www.ibtimes.com/mobile-phones-africa-subscriber-growth-slow-sharply-companies-struggle-reach-rural-2140044

[48] Fox, K., "Africa's Mobile Economic Revolution", *The Guardian*, July 23, 2011, accessed at http://www.theguardian.com/technology/2011/jul/24/mobile-phones-africa-microfinance-farming

[49] Aker, J., Mbiti, I., "Mobile Phones and Economic Development in Africa", *Journal of Economic Perspectives*, Volume 24, Number 3, Pages 207–232, Summer 2010.

[50] McQuail, D., *Media Accountability and Freedom of Publication*, Oxford University Press, November 20, 2003.

[51] "Social Media as a Credible News Source", *New York Women in Communications*, 2013, accessed at http://www.nywici.org/features/social-media-credibility

[52] Pouster J., et. al., "Internet Seen as Positive Influence on Education but Negative On Morality in Emerging and Developing Nations", Pew Research Center, March 19, 2015, accessed at http://www.pewglobal.org/files/2015/03/Pew-Research-Center-Technology-Report-FINAL-March-19-20151.pdf

[53] Cerf, V.G., "Internet Access Is Not a Human Right", *The New York Times*, January 4, 2012, accessed at http://www.nytimes.com/2012/01/05/opinion/internet-access-is-not-a-human-right.html

[54] "Arab Spring", Wikipedia, accessed at https://en.wikipedia.org/wiki/Arab_Spring

[55] Keller, J., "Evaluating Iran's Twitter Revolution", *The Atlantic*, June 18, 2010, accessed at

http://www.theatlantic.com/technology/archive/2010/06/evaluating-irans-twitter-revolution/58337/

[56] Rigg. J., "China bans news sites from using social media as a source", Engadget, July 4, 2016, accessed at https://www.engadget.com/2016/07/04/china-bans-news-sites-from-using-social-media-as-a-source/

[57] Oster, S., "China Fakes 488 Million Social Media Posts a Year: Study", Bloomberg, May 19, 2016, accessed at http://www.bloomberg.com/news/articles/2016-05-19/china-seen-faking-488-million-internet-posts-to-divert-criticism

[58] Gessen, M., "How Putin Controls the Internet and Popular Opinion in Russia", The Intercept, September 8, 2015, accessed at https://theintercept.com/2015/09/08/how-putin-controls-the-russian-internet/

[59] Berlatsky, N., "Hashtag Activism Isn't a Cop-Out", *The Atlantic*, January 7, 2015, accessed at http://www.theatlantic.com/politics/archive/2015/01/not-just-hashtag-activism-why-social-media-matters-to-protestors/384215/

[60] Phillips, C., "Who is watching the hate? Tracking Hate Groups Online and Beyond", Public Broadcasting Service, April 1, 2016, accessed at http://www.pbs.org/independentlens/blog/who-is-watching-the-hate-tracking-hate-groups-online-and-beyond

[61] Fernanado, G., "Inside the 'Black Separatist Movement', the Radical Response to Racism in America", News.com.au, July 13, 2016, accessed at http://www.news.com.au/world/north-america/inside-the-black-separatist-movement-the-radical-response-to-racism-in-america/news-story/d7e5e54d1de7cc3df16b84626c968e31

[62] Dickson, C., "Where White Supremacists Breed Online", The Daily Beast, April 17, 2014, accessed at http://www.thedailybeast.com/articles/2014/04/17/where-white-supremacists-breed-online.html

[63] Chrisafis, A., "French Police Chief and partner killed in stabbing claimed by Isis", *The Guardian*, June 14, 2016, accessed at https://www.theguardian.com/world/2016/jun/13/french-policeman-stabbed-death-paris

[64] Purcell, K., Rainie, L., "Americans Feel Better Informed Thanks to the Internet", Pew Research Center, December 8, 2014, accessed at http://www.pewinternet.org/2014/12/08/better-informed/

[65] Sahlin, J. P., *Social Media and the Transformation of Interaction in Society*, IGI Global, September 2015.

[66] Santana, M., "Orlando Authorities Use Social Media to Inform Public on Shootings", Emergency Management, June 12, 2016, accessed at http://www.emergencymgmt.com/safety/Orlando-authorities-use-social-media-to-inform-public-on-shootings.html

[67] "Social Media Aid the Haiti Relief", Pew Research Center: Journalism and Media, January 21, 2010, accessed at http://www.journalism.org/2010/01/21/social-media-aid-haiti-relief-effort/

---

[68] "Monsanto at a Glance", Monsanto, 2015, accessed at http://www.monsanto.com/whoweare/pages/default.aspx
[69] "Monsanto's Genetically Engineered Wheat Scandal Is No Surprise", *Forbes*, June 5, 2013, accessed at http://www.forbes.com/sites/forbesleadershipforum/2013/06/05/monsantos-genetically-engineered-wheat-scandal-is-no-surprise/
[70] "Wheat Improvement: The Truth Unveiled", United States Department of Agriculture, accessed at http://wheat.pw.usda.gov/ggpages/Wheat__Improvement-Myth_Versus_FactFINAL.pdf
[71] Giminez, E.H., "We Already Grow Enough Food for 10 Billion – and Still Can't End Hunger", The Huffington Post, May 2, 2012, accessed at http://www.huffingtonpost.com/eric-holt-gimenez/world-hunger_b_1463429.html
[72] "Meat and Animal Feed", Foundation on Future Farming, accessed at http://www.globalagriculture.org/report-topics/meat-and-animal-feed.html
[73] Ibid.
[74] "Global Food Losses and Food Waste – Extent, Causes and Prevention", Food And Agriculture Organization of the United Nations, 2011, accessed at http://www.fao.org/docrep/014/mb060e/mb060e.pdf
[75] Ibid.
[76] "A Better Way to Connect Food and People", Evaptainers, 2016, accessed at http://www.evaptainers.com
[77] "Chemical Engineering Innovation in Food Production", Chemical Engineers in Action, accessed at http://www.chemicalengineering.org/docs/cheme-food.pdf
[78] "Food Additive Status List", Food and Drug Administration, December 2014, accessed at http://www.fda.gov/Food/IngredientsPackagingLabeling/FoodAdditivesIngredients/ucm091048.htm
[79] Smithers, R., "Almost half of the world's food thrown away, reports finds", *The Guardian*, January 10, 2013, accessed at http://www.theguardian.com/environment/2013/jan/10/half-world-food-waste
[80] "Facts", End Food Waste Now, 2013, accessed at http://www.endfoodwastenow.org/index.php/resources/facts
[81] "Hunger and Poverty Facts and Statistics", Feeding America, accessed at http://www.feedingamerica.org/hunger-in-america/impact-of-hunger/hunger-and-poverty/hunger-and-poverty-fact-sheet.html
[82] "Hunger Statistics", United Nations World Food Programme, 2016, accessed at https://www.wfp.org/hunger/stats
[83] "Facts on the Fast Food Industry", Statistica, accessed at http://www.statista.com/topics/863/fast-food/
[84] "Trends in International Trade", World Trade Organization, 2013, accessed at https://www.wto.org/english/res_e/booksp_e/wtr13-2b_e.pdf
[85] New, C., "Made In America Is A Luxury Label That Will Cost You", The Huffington Post, 9/17/2012, accessed at

http://www.huffingtonpost.com/2012/09/17/made-in-america-the-luxury-label-will-cost-you_n_1891127.html

[86] Fox, K., "Africa's Mobile Economic Revolution", *The Guardian*, July 23, 2011, accessed at http://www.theguardian.com/technology/2011/jul/24/mobile-phones-africa-microfinance-farming

[87] "India Country Summary of Higher Education", The World Bank, accessed at http://siteresources.worldbank.org/EDUCATION/Resources/278200-1121703274255/1439264-1193249163062/India_CountrySummary.pdf

[88] "State of Education in Africa Report 2015", The Africa-America Institute, 2015, accessed at http://www.aaionline.org/wp-content/uploads/2015/09/AAI-SOE-report-2015-final.pdf

[89] Michael, R., "Education in China", World Education Services, March 7, 2016, accessed at http://wenr.wes.org/2016/03/education-in-china-2/

[90] "Girls Education Plays a Large part in Global Development", United Nations Girls' Education Initiative, accessed at http://www.ungei.org/news/247_2165.html

[91] Lynch, M., "4 Ways Digital Tech has Changed K-12 Learning", *The Journal*, May 20, 2015, accessed at https://thejournal.com/articles/2015/05/20/4-ways-digital-tech-has-changed-k12-learning.aspx

[92] "How has Technology Changed Education", Purdue University, 2016, accessed at http://online.purdue.edu/ldt/learning-design-technology/resources/how-has-technology-changed-education

[93] Tomaszewski, J., "Do Texting and Facebook belong in the Classroom", Education World, 2011, accessed at http://www.educationworld.com/a_tech/archives/technology/social_media_in_classroom.shtml

[94] Michaels, D., *Doubt is Their Product: How Industry's Assault on Science Threatens Your Health,* Oxford University Press, 2008.

[95] "Assistive Technology devices", Public Broadcasting Service, accessed at http://www.pbs.org/parents/education/learning-disabilities/strategies-for-learning-disabilities/assistive-technology-devices/

[96] "The Positive Impact of eLearning – 2012 Update", United Nations Educational, Scientific and Cultural Organization, 2012, accessed at http://www.unesco.org/new/fileadmin/MULTIMEDIA/HQ/ED/pdf/The%20Positive%20Impact%20of%20eLearning%202012UPDATE_2%206%20121%20(2).pdf

[97] Highley, M., "e-Learning: Challenges and Solutions", eLearning Industry, March 14, 2014, accessed at https://elearningindustry.com/e-learning-challenges-and-solutions

[98] Andrews, D.J.C., Bartell, T., Richmond, G., "Teaching in Dehumanizing Times, the Professional Imperative", *Journal of Teacher Education*, Vol. 67 No. 3, pp. 170-172, May/June 2016.

[99] "Massive Open Online Courses (MOOCS) Directory", AboutEdu Inc., 2016, accessed at http://www.moocs.co/Higher_Education_MOOCs.html

[100] Coursera, 2016, accessed at https://www.coursera.org/

[101] "New study on MOOCs in developing countries reveals half of users receive certification", IREX, April 11, 2016, accessed at https://www.irex.org/news/new-study-moocs-developing-countries-reveals-half-users-receive-certification
[102] "International Students in the United States", Institute of International Education, 2016, accessed at http://www.iie.org/Services/Project-Atlas/United-States/International-Students-In-US
[103] "Open Doors Data – U.S. Study Abroad", Institute of International Education, 2016, accessed at http://www.iie.org/Research-and-Publications/Open-Doors/Data/US-Study-Abroad
[104] Ruiz, N.G., "The Geography of Foreign Students in U.S. Higher Education: Origins and Destinations", The Brookings Institute, August 29, 2014, accessed at http://www.brookings.edu/research/interactives/2014/geography-of-foreign-students#/M10420
[105] Anand, D.K., Frehill, L., Hazelwood, D., Kavetsky, R., and Ryan, E., *S&E Revitalization: A New Look,* CALCE EPSC Press, Center for Engineering Concepts Development, University of Maryland, College Park, MD, 2012.
[106] Stem Education Coalition, 2016, accessed at http://www.stemedcoalition.org/
[107] "Today in Energy", US Energy Information Administration, 2016, accessed at http://www.eia.gov/todayinenergy/detail.cfm?id=11951
[108] "U.S. Coal and the U.S. Economy", 2016, Federation for American Coal, Energy and Security, accessed at http://www.facesofcoal.org/index.php?u-s-coal-and-the-u-s-economy
[109] Colagiuri, R., Cochrane, J., Girgis, S., "Health and Social Harms of Coal Mining In Local Communities", Beyond Zero Emissions, October 2012, accessed at, https://sydney.edu.au/medicine/research/units/boden/PDF_Mining_Report_FINAL_October_2012.pdf
[110] Munnik, V., "The Social and Environmental Consequences of Coal Mining in South Africa", Both Ends, January, 2010, accessed at http://www.bothends.org/uploaded_files/uploadlibraryitem/1case_study_South_Africa_updated.pdf
[111] Friedman, L., "Coal-Fired Power in India May Cause More Than 100,000 Premature Deaths Annually", *Scientific American*, March 11, 2013, accessed at http://www.scientificamerican.com/article/coal-fired-power-in-india-may-cause-more-than-100000-premature-deaths-annually/
[112] Mudway I.S., et al., "Combustion of dried animal dung as biofuel results in the generation of highly redox active fine particulates", *Part Fibre Toxicol.* Vol. 2, Issue 6, October 2005.
[113] Stumpf, J., "Climate impacts of kerosene lamps used in developing countries", National Institutes of Health, accessed at http://www.niehs.nih.gov/news/newsletter/2013/1/science-kerosene/
[114] "Energy for Cooking In Developing Countries", International Energy Agency, accessed at https://www.iea.org/publications/freepublications/publication/cooking.pdf

[115] "Modi's 2022 Renewable Energy Target Requires Four Times as Much Money as Defence Budget", Fact Checker, December 9, 2015, accessed at http://factchecker.in/modis-175-gw-renewable-energy-target-for-2022-needs-160-billion-investment/

[116] "2015 Key World Energy Statistics", International Energy Agency, 2015, accessed at http://www.iea.org/publications/freepublications/publication/KeyWorld_Statistics_2015.pdf

[117] Ibid.

[118] Canning P., et al., "Energy Use in the U.S. Food System", United States Department of Agriculture, accessed at http://www.ers.usda.gov/media/136418/err94_1_.pdf

[119] "How we use Energy - Industry", The National Academy of Sciences, 2016, accessed at http://needtoknow.nas.edu/energy/energy-use/industry/

[120] "How we use Energy - Transportation", The National Academy of Sciences, 2016, accessed at http://needtoknow.nas.edu/energy/energy-use/transportation/

[121] Davis, L.W., Gertler, P.J., "Contribution of air conditioning adoption to future energy use under global warming", *Proceedings of the National Academy of Sciences of the United States of America*, December 9, 2014, accessed at http://www.pnas.org/content/112/19/5962.full.pdf

[122] Sivak, M., "Will AC Put a Chill on the Global Energy Supply?", *American Scientist*, 2014, accessed at http://www.americanscientist.org/issues/pub/will-ac-put-a-chill-on-the-global-energy-supply

[123] Chen, F., "Why Hongkongers can't live without air-conditioning", *Hong Kong Economic Journal*, October 24, 2015, accessed at http://www.ejinsight.com/20151023-why-hongkongers-can-t-live-without-air-conditioning/

[124] "The Future of Air Conditioning", Children's Investment Fund Foundation, accessed at https://ciff.org/news/future-air-conditioning/

[125] Davis, L.W., Gertler, P.J., "Contribution of air conditioning adoption to future energy use under global warming", *Proceedings of the National Academy of Sciences of the United States of America*, December 9, 2014, accessed at http://www.pnas.org/content/112/19/5962.full.pdf

[126] Tovey, A., "Nissan unveils new power system where electric cars feed energy back into the grid", *The Telegraph*, May 10, 2016, accessed at http://www.telegraph.co.uk/business/2016/05/10/nissan-unveils-new-power-system-where-electric-cars-feed-energy/

[127] Farahani, F., "Interface of Cultural Identity Development", Indira Gandhi National Centre for the Arts, accessed at http://ignca.nic.in/ls_03019.htm

[128] Schoettle, B., Sivak, M., "A Survey of Public Opinion about Autonomous and Self-Driving Vehicles in the U.S., the U.K., And Australia", University of Michigan, July, 2014, accessed at https://deepblue.lib.umich.edu/bitstream/handle/2027.42/108384/103024.pdf

[129] Sullivan, M., "America's most tech-friendly cities", PC World, December 20, 2012, accessed at http://www.pcworld.com/article/2022294/america-s-most-tech-friendly-cities.html

[130] Bertasio, D., "The Role of Culture in the Technological Advancement Process", *D. AI & Soc*, Volume 7, Issue 3, pp. 248-252, Springer, September 1993.

[131] Kuchikomi, "National obsession with cleanliness bodes ill for health", Japan Today, April 30, 2013, accessed at http://www.japantoday.com/category/kuchikomi/view/national-obsession-with-cleanliness-bodes-ill-for-health

[132] Fifield, A., "How Japan's toilet obsession produced some of the world's best bathrooms", *The Washington Post,* December 15, 2015, accessed at https://www.washingtonpost.com/news/worldviews/wp/2015/12/15/how-japans-toilet-obsession-produced-some-of-the-worlds-best-bathrooms/

[133] "Very clean people, the Japanese", *The Economist*, July 31, 1997, accessed at http://www.economist.com/node/153179

[134] Mims C., "Why Japanese Love Robots (And Americans Fear Them)", *MIT Technology Review,* October 12, 2010, accessed at https://www.technologyreview.com/s/421187/why-japanese-love-robots-and-americans-fear-them/

[135] "What is the Amish Ordnung?", Amish America, 2010, accessed at http://amishamerica.com/what-is-the-amish-ordnung/

[136] "Technology", Elizabethtown College, 2016, accessed at https://groups.etown.edu/amishstudies/cultural-practices/technology/

[137] Bertasio, D., "The Role of Culture in the Technological Advancement Process", *D. AI & Soc,* Volume 7, Issue 3, pp. 248-252, Springer, September 1993.

[138] "Foreign Direct Investment", India Brand Equity Foundation, 2016, accessed at http://www.ibef.org/economy/foreign-direct-investment.aspx

[139] "India census says 70 percent live in villages, most are poor", *The Seattle Times*, July 3, 2015, accessed at http://www.seattletimes.com/nation-world/world/india-census-says-70-percent-live-in-villages-most-are-poor/

[140] "Now, you can do a virtual 'puja' for your favourite Ganpati in Mumbai", NDTV Convergence Limited, September 23, 2012, accessed at http://www.ndtv.com/india-news/now-you-can-do-a-virtual-puja-for-your-favourite-ganpati-in-mumbai-499982

[141] "Indian village where women have been banned from using mobile phones because 'they could use them to elope'", *The Daily Mail*, December 6, 2012, accessed at http://www.dailymail.co.uk/news/article-2243481/Indian-women-banned-using-mobile-phones-council-claims-encourage-to.html

[142] Mohan, V., "Gujarat Village Diktat: No Mobile for Unmarried Women", The Huffington Post, July 15, 2016, accessed at http://www.huffingtonpost.in/2016/02/19/modi-gujarat-village-bans_n_9272882.html

---

[143] "Amendment II Right to Bear Arms", National Constitution Center, accessed at http://constitutioncenter.org/interactive-constitution/amendments/amendment-ii

[144] "You can't carry a tennis ball near the GOP convention. But you can bring a gun.", *The Washington Post*, July 14, 2016, accessed at https://www.washingtonpost.com/opinions/you-cant-carry-a-tennis-ball-near-the-gop-convention-but-you-can-bring-a-gun/2016/07/14/8216a202-4932-11e6-90a8-fb84201e0645_story.html

[145] "Number of guns per capita by country", Wikipedia, accessed at https://en.wikipedia.org/wiki/Number_of_guns_per_capita_by_country

[146] Wilmore, J., "America's gun industry is booming", Salon, December 3, 2012, accessed at http://www.salon.com/2012/12/03/americas_gun_industry_is_booming/

[147] "Arms Industry", Wikipedia, accessed at https://en.wikipedia.org/wiki/Arms_industry

[148] Xu, J., et al., "National Vital Statistics Reports", Centers for Disease Control and Prevention, February 16, 2016, accessed at http://www.cdc.gov/nchs/data/nvsr/nvsr64/nvsr64_02.pdf

[149] Lawrence L.C., "Why You Can't Buy a Smart Gun", Center for American Progress Action Fund, January 8, 2016, accessed at http://thinkprogress.org/justice/2016/01/08/3736523/smart-gun-obstacles/

[150] Ibid.

[151] "TED: The Economics Daily image", Bureau of Labor Statistics, October 21, 2014, accessed at http://www.bls.gov/opub/ted/2014/ted_20141021.htm

[152] McKeever, B., "The Nonprofit Sector in Brief 2015: Public Charities, Giving, and Volunteering", The Urban Institute, October 29, 2015, accessed at http://www.urban.org/research/publication/nonprofit-sector-brief-2015-public-charities-giving-and-volunteering

[153] "About Us", Engineers Without Borders USA, 2016, accessed at http://www.ewb-usa.org/our-story/about-us/

[154] "Approach", Engineers Without Borders USA, 2016, accessed at http://www.ewb-usa.org/our-story/approach/

[155] Engineering for Change, 2016, accessed at http://www.engineeringforchange.org/

[156] "Who We Are", USAID, 2016, accessed at https://www.usaid.gov/who-we-are

[157] "Higher Education Solutions Network (HESN)", USAID, 2016, accessed at https://www.usaid.gov/hesn

[158] "What is GEF", Global Environment Facility, 2016, accessed at https://www.thegef.org/gef/whatisgef

[159] "GEF Projects", Global Environment Facility, 2016, accessed at https://www.thegef.org/gef/gef_projects_funding

[160] "A world of development experience", United Nations Development Programme, 2016, http://www.undp.org/content/undp/en/home/operations/about_us.html

---

[161] "Open Projects", United Nations Development Programme, 2016, accessed at http://open.undp.org/#2016

[162] "Leading Corporate Philanthropy", Council on Foundations, 2016, http://www.cof.org/content/leading-corporate-philanthropy

[163] "The Corporate Foundation Advantage Plan", The American Foundation, 2016, accessed at http://www.americanfoundation.org/corporate-foundations/

[164] "Intel Supporting Local Schools, Non-Profits and Education", Intel Corporation, 2016, accessed at http://www.intel.com/content/www/us/en/corporate-responsibility/intel-invests-in-our-communities.html

[165] Elam D.A., "A letter from Deb Elam", General Electric Foundation, accessed at http://www.gefoundation.com/about-ge-foundation/a-letter-from-deb-elam/

[166] "Worldwide Giving", Exxon Mobil, 2016, accessed at http://corporate.exxonmobil.com/en/community/worldwide-giving/exxonmobil-foundation/overview

[167] "Q&A: What is a Corporate Foundation?", *The Guardian*, June 13, 2007, accessed at http://www.theguardian.com/society/2007/jun/13/societyguardian.societyguardian2

[168] "What We Do", The Bill and Melinda Gates Foundation, 2016, accessed at http://www.gatesfoundation.org/What-We-Do

[169] "U of T Engineers Put Their Heads Together to Reinvent the Toilet", Bill and Melinda Gates Foundation, 2016, accessed at http://www.gatesfoundation.org/Media-Center/Press-Releases/2011/07/U-of-T-Engineers-Put-Their-Heads-Together-to-Reinvent-the-Toilet

[170] Gates, B. and Gates, M., "Two Superpowers We Wish We Had", The Bill and Melinda Gates Foundation, 2016, accessed at https://www.gatesnotes.com/2016-Annual-Letter

A solar engineer in Legga village, Rajasthan, India [171]

A member of "Serve On" holds up a flying drone used to help identify areas that were worst-hit by the April 2015 earthquake in Nepal [172]

Chapter 3

Unintended Consequences

It has become appallingly obvious that our technology has exceeded our humanity.[173]

In early 2016, the contents of an Apple iPhone belonging to one of the terrorists involved in the killing of 14 people in San Bernardino, California, were seemingly inaccessible to government investigators. Apple refused the government's request to unlock it. The phone's security measures were later defeated by an outside company at considerable expense to the government, but it brought to the fore a public discussion of the rights of the various actors in relation to technology and security. Following the resolution of this particular case, it was later reported with little fanfare that Apple would now be working on encrypting its devices such that even if compelled by the government, no action could be taken to effectively unlock them. Whether this is a positive or negative development depends upon whom and when you ask. This is an example of an unintended consequence of our improvements in engineered technologies that raise questions of whether we should create technology that we may not be able to control.

It is not difficult to observe the unintended consequences of our often unrestrained technological and engineering development over the past century. In fact, many would argue that even glancing outside in many places of the world highlights a wide variety of issues. Environmental pollution, waste, and health problems are some obvious examples. While not comprehensive, we will profile some important unintended consequences worthy of consideration.

3.1 Environmental Pollution

The effect of engineering on the environment has been nothing short of extraordinary. Humanity has carved up the earth in almost every way possible. Entire mountains have been leveled, forests erased, even parts of entire oceans poisoned - the earth has been plundered for the improvement and enrichment of human existence. We have recently discovered that human pollutants permeate the ecosystem even at the very bottom of our oceans. In the Marianas Trench, in the Western Pacific Ocean, and at a depth of 11 kilometers, high levels of PCBs, or polychlorinated biphenyls (industrial chemicals) were found. They were at levels "even higher than in the estuaries of two of the most polluted rivers — the Pearl River and the Liao River — in China." [174]

A common problem for tens of millions of homes, mostly those built before 1986, is water pollution caused by lead which is leached into drinking water as it comes into individual properties. Lead pipes were once considered an appropriate choice for water plumbing due to the metal's malleability to form the shapes required, and later a lead-tin alloy solder was widely used to join copper plumbing pipe. Long-term health problems eventually moved the industry away from lead-based piping, but it remains in place in many locations throughout the world, even in the US. When coated appropriately, or when transporting water that is alkaline, there are few health concerns with lead piping. However, as in the case of the lead poisoning cases in Flint, Michigan, if the water being pumped through is acidic, then lead is leached out into drinking water and consumed. Residents of the city believed the water was safe to drink after previous government-mandated testing of the water supply found acceptable lead levels, whereas independent testing verified high levels of lead contamination.

A decade ago, administrators in the public school system in Baltimore, Maryland, discovered the same problem as water pipes throughout their 180 schools were delivering unsafe levels of lead for students to drink. Without the funds for the estimated millions of dollars per school required to tear out the old pipes and replace them, they took the unusual but cost-effective route of spending $450,000 per year to provide bottled drinking water to all schools.[175] This effectively skirted the issue and kept children safe from the risk of contaminated water at the cost of a lot of wasted energy and time transporting what should be a dependable resource in a developed country.

These issues are not just limited to specific communities like Flint, Michigan, or Washington, D.C., or to any of the high profile lead contamination stories from the past decade. In one organization's testing, unacceptable lead levels were found in the water supply in almost 2,000 water systems in the US.[176] Currently, the only engineering solution to this widespread threat to human health is the treatment of drinking water with anti-corrosion chemicals, a quick fix at best that attempts to simply slow the rate of lead contamination in pipes but introduces yet another chemical into the water supply.

Another form of water pollution is the result of our success in developing and delivering more effective fertilizers, used in a variety of venues from mega-farms all the way down to suburban backyards. When rains wash through recently fertilized land they can carry the fertilizer out of the soil and into waterways, where the excess nitrates fuel the growth of algae that chokes underwater life and stagnates water by removing oxygen. In an examination of this growing problem, *Scientific American* found that our increased production of ethanol biofuel from corn is contributing to the problem. This is a secondary unintended consequence of our engineering success in efficient large scale ethanol fuel production.

> *What is clear is that a significant portion of such fertilizer is still making its way through the soil and water to the sea. As a result, algae and other microorganisms take up the nitrogen, bloom and, after they die, suck the oxygen out of coastal waters. Such "dead zones" have appeared seasonally near most major river mouths, including those emptying into Maryland's Chesapeake Bay as well as the Gulf of Mexico, where lifeless waters now cover more than 7,700 square miles (20,000 square kilometers) during the summer months.*[177]

Public policy has played a role in attempting to mitigate the problem with the passing of laws such as the State of Maryland's Fertilizer Use Act. This law restricts the use of fertilizers to certain concentrations and certain times of the year. This affects the one million acres of managed grass areas in the state, almost equal to the amount of farmland.[178] While a step in the right direction, some believe this problem is advanced enough that it is unlikely to be effectively addressed with legislation. Our society must reduce its dependence on green lawns and rapid produce growth, or we must engineer replacement fertilizers in order to reduce the danger to our water supplies. "In California, the richest agricultural area of the richest country in the world, the impoverished farmworkers on whom the industry depends must also endure drinking water contaminated by agricultural chemicals."[179]

Our volume of trash has risen exponentially over the past fifty years as our modern lifestyle utilizes more and more packaging. *The Atlantic* observed in 2012 that the average person in a developed country, like the US, is producing 2.6 pounds of garbage every day, with an overall total of 2.6 trillion pounds being produced globally.[180] Higher-income Organization for Economic Cooperation and Development (OECD) member countries make up half of this volume, with only 1% composted and 1% recycled. In lower-income countries, less than 1% is recycled, and 1% is composted. This leaves the overwhelming majority of all trash globally ending up in landfills (59%), dumps (13-33%), or other locations without being reused or recycled. With so much trash entering landfills, it is important also to note that unless it is carefully constructed using modern waste engineering techniques, groundwater is often contaminated with leachate, which can then poison drinking water and threaten the health of local communities.[181]

A subsurface smoldering fire beneath a trash pile is an unexpected danger of our buried trash. This is the case in the Bridgeton Landfill, 20 miles from St. Louis, Missouri. The process of decomposition typically heats the interior of a landfill to around 140 degrees Fahrenheit, but "parts of the Bridgeton landfill, in contrast, have reached temperatures as high as 260. That 120 degrees is the difference between a healthy landfill, decomposing merrily along, and one in which the systems of safe waste management are falling apart."[182] The landfill was recently discovered to be much closer than originally thought, at only a thousand feet, from the site of radioactive waste material from the Manhattan Project, a research and development project that created nuclear weapons during World War II. After six years, plans are still underway to try to engineer an effective long-term solution to prevent the radioactive waste from becoming airborne. Currently, a temporary solution is being used where a makeshift radiator has been designed and installed. This fix involves using water pipes that have been sunk into the ground, with cool water flushed through them to draw heat out of the ground, at a cost in the tens of millions of dollars.

The subsurface smoldering waste fire is not just a problem in the US, and in fact is much greater in some other countries. Since opening in 1927, Mumbai's Deonar garbage dump has grown into a 300-acre space where more than 5,000 tons of garbage are dumped each day without sorting or management. During 2016 the dump caught fire multiple times, while residents suffering from exposure to the toxic smoke emanating from the dump have tried to fight to have the city relocate the dump.[183] The failure of public policy to protect citizens in the area is surprising from a developed world perspective. The city does not issue health advisories, even with infants dying of respiratory illnesses. Parents keep children home from school to avoid the danger, and health clinics are seeing a greater number of patients for respiratory problems due to the toxic air quality. And yet, despite the obvious danger, some protest shutting down or moving the dump:

> *In a large slum that hugs the dump, children playing cricket run in and out of the area to fetch the ball. Rotting trash overflows the dump, floating in open drains and collecting in small heaps outside homes.*

> *Infants cough all day, while billboards hanging overhead demand that the dump be shut down. And yet, said Syed Shah, 46, a trash picker whose family of seven lives less than 20 feet from the noxious mountain, "this garbage dump is our daily bread." He said he feeds his family by selling the plastic and metal scrap and coconut shells that he pulls from the dump each day.* [184]

Some of the engineered materials we have created, such as plastics, have truly revolutionized the way we live our lives. Since plastic's creation, however, there has never been a sustained large-scale effort to change the way we dispose of it. Nor have we developed low cost chemical variations that allow it to quickly break down into a harmless form. Consequently, we are beginning to pay the price for our tremendous success with plastic. While a wonder for packaging, plastics have been and continue to be an environmental disaster of epic proportions, both on land and in the oceans. A recent estimate by the Ocean Recovery Alliance was that 33% of the plastic used in the world is used once then discarded, with only 15% worldwide being recycled. An ocean advocacy group, 5 Gyres Institute, undertook a comprehensive four year global survey of ocean trash and estimated that 5.25 trillion pieces of trash are currently on the surface of the world's oceans.[185] A study in the journal *Science* estimated that globally eight million metric tons of plastics end up in our oceans each year. This equates to "five plastic grocery bags filled with plastic for every foot of coastline in the world," a truly staggering number that by 2025 is projected to double.[186]

Microbeads, or extremely small plastic beads most often used in soaps, body scrubs and toothpaste, are a very effective abrasive. As such, they were quickly integrated into many products on supermarket shelves. As has been the case with plastics in years past, these beads were strong enough to not break down for decades. They also suffer from two additional environmental problems, the first being that they are too fine to be effectively filtered out of the water system, and the second that they often bind with pesticides before going out to the sea. This results in what the director of an environmental group termed a "toxic pill."[187] Mistaking the beads for food, fish eat the plastic beads, and then the same fish are caught and served back to humans. It was estimated in 2015 that every year, just in the state of New York, 19 tons of microbeads were entering waterways. The Microbead-Free Waters Act of 2015 has now legislated that all companies must start phasing out using microbeads in their products by July 2017, with a total ban by 2020.[188]

While this is a positive step forward in policy, microbeads are not the only source of unfilterable plastic particles in our oceans and other waterways. Regular plastics also break down into very small particles as they age, as observed in the Great Pacific Garbage Patch, a large and fairly stationary region in the Pacific Ocean where plastic pollution is concentrated by ocean currents. The region consists of plastic waste both in larger forms on the surface and also broken down to microplastic size in the upper layers of the water. Estimates of the size of the Great Pacific Garbage Patch range from 0.4% to 8% of the Pacific Ocean, and much of the debris is small and subsurface, making for difficult observation.[189]

Improved waste management infrastructure is considered to be the key to reducing plastic pollution. An illustrative comparison is the fact that up to 3.5 million metric tons of plastic waste enters the ocean from China's coastal cities each year, while only 110,000 metric tons come from the US due to better waste management practices.[190] The combined role of public policy and engineering is clear here in addressing the problem – use less plastic in packaging, build better waste management systems, and create public policy to support these efforts.

Electronic waste (e-waste) is an unintended consequence of a combination of Moore's Law, our engineering prowess, and our insatiable desire for faster, newer electronics with more features. Our ability to produce new devices with greater functionality at a reasonably affordable price along with a healthy economy and high wages, encourage a mentality of waste. As engineering has enabled more efficient production, more flexible materials and resultant cheaper products, the technology-enabled developed world has adopted the approach of "Doesn't work? Throw it out, buy a new one!" As noted by Scott Nealy, former Chairman and CEO of Sun Microsystems, "Technology has the shelf life of a banana, especially in the hardware space."[191]

Electronics are one of our most dangerous forms of waste, as they contain toxic materials such as lead, cadmium, mercury, and arsenic. It is also our fastest growing type of waste, with estimates in 2013 that 50 million tons of e-waste worldwide are created every year. Only 15-20% is estimated to have been recycled, with developing countries becoming dumping grounds.[192] In a study tracking e-waste from the US, it was found that often it is exported internationally instead of being recycled in the US. The authors of the study tracked items of e-waste using GPS trackers hidden inside the electronics, and discovered a third of the tracked items ended up being exported to Mexico, Taiwan, China, Pakistan, Thailand, the Dominican Republic, Canada, and Kenya. Much of the e-waste exported from the United States ends up in China, where recycling and recovery processes are undertaken with little enforcement of safety laws and little to no protection of workers or the environment. Recycling methods are very crude in these regions, with workers desoldering circuit boards over fires, soaking chips in acid to recover gold, silver and platinum, burning off plastic wire casings, and ultimately polluting water sources and the air with cancer-causing toxins.[193]

Engineering improvements in mining technology are having a positive impact on society with cheaper natural gas prices and greater domestic supplies, but in the US we are also dealing with unintended consequences of our improved ability to extract these materials. Fracking, a process of drilling into the earth and injecting a mix of water, sand and chemicals to release gas from the ground, is both chaotic and dangerous. "Induced" earthquakes are a result of wastewater injection into the ground and occur more frequently than ever before due to increased fracking operations in the US. There were approximately 29 magnitude 3+ earthquakes per year from 1970 to 2009 in the central and eastern regions of the US. As fracking operations increased over the past decade, the number climbed exponentially to 659 in 2014. Necessarily, the United States Geological Survey (USGS) released a new hazard model in 2016 for induced earthquake prediction.[194] Another unintended consequence of fracking operations is referred to as methane migration, where gas contaminates underground water wells due to poor mining practices. This has resulted in drinking water that catches on fire due to small pockets of methane that are trapped in the water.[195]

Our global need for toxic materials is poisoning the places where these materials are deposited and mined due to poorly regulated mining practices. In countries where technically sound public policy has not been carefully and aggressively implemented and enforced, a significant human and environmental toll has been paid. China's rare earth materials mining and processing industry serves as a clear example of this problem. When rare earth materials such as neodymium were needed in large quantities for the manufacturing of high technology items such as hybrid vehicles, prices for the materials escalated. Chinese companies mined them in volume using techniques that caused tremendous environmental and health problems for those living in nearby areas. As noted in the book *Rare Earth Materials: Insights and Concerns*, producers in the Baotou region discharged approximately 10 million tons of contaminated wastewater every year as part of the mining process, poisoning farmland and local water resources. Some of the wastewater is also radioactive and contains significant quantities of acids used to leach the rare earth materials from mined rock.[196]

According to the World Health Organization (WHO), close to 3 billion people still use wood, animal dung, coal, and crop waste for cooking and heating. Burning biomass fuels such as these (and in some cases, even trash plastics) has a very serious impact on long-term health. Over 4 million people die each year globally from illnesses related to burning these fuels, and more than 50% of pneumonia deaths in children under 5 in developing countries result from pollution from these fires inside the home.[197] In many cases there are simple engineering fixes for the inefficient cooking stoves used in millions of homes, but even low cost replacement options are out of reach for many. It is extraordinary that those who are using kerosene lamps in developing countries represent almost four times the entire population of the US. This segment of the global population is so constrained by their lack of electricity and lack of funds for purchasing better choices, (solar panel driven LED lights, for instance), that they are being poisoned by the only option they have.

An interesting and subtle form of pollution has impacted our ability to observe and enjoy the beautiful sky at night. According to imagery released in June 2016 by the National Oceanic and Atmospheric Administration (NOAA), one third of humanity and up to 80% of Americans can no longer see the Milky Way in the sky at night due to light pollution. Chris Elvidge, a scientist with NOAA's National Centers for Environmental Information in Boulder, Colorado, states that "We've got whole generations of people in the United States who have never seen the Milky Way. It's a big part of our connection to the cosmos - and it's been lost."[198]

3.2 Internet Misuse

In 2014, it was estimated by the FBI's Internet Crime Complaint Center (IC3) estimated that losses in the US from cybercrime totaled over $800 million from 269,422 complaints.[199] The organization believes that only 10% of cybercrimes are reported to the IC3, making the actual figure in the billions. This crime took many forms:

- Government impersonation email scams
- Intimidation/extortion scams
- Real estate fraud
- Auto fraud
- Confidence fraud/romance scams
- Virtual currency schemes
- Business email compromises

So far, we have not successfully engineered an effective solution to this problem. While acknowledging that this is typically a social engineering problem, some say it is an unintended consequence of the flawed design of the Internet itself. "It's not that we didn't think about security," said MIT scientist David Clark, one of the early developers of the Internet. "We knew that there were untrustworthy people out there, and we thought we could exclude them." Vinton G. Cerf, another early Internet developer, further notes that "We didn't focus on how you could wreck this system intentionally....You could argue with hindsight that we should have, but getting this thing to work at all was non-trivial."[200]

The use of the Internet to anonymously convey immoral, illegal or undesirable content contributes negatively to society's ills. This is particularly true for the young and impressionable, who now include their online presence in the development of their overall persona. There are thousands of news stories of online bullying, misogyny, child pornography, and harassment that result in shattered lives and even suicides and homicides. With the combination of the "Wild West" of unregulated speech on the Internet, a herd mentality that encourages negativity against those who may be "different," and few controls in place to stop young people from using and misusing a dizzying array of communications services, it is of little surprise that a terrible fate befalls many young people every year. As the parent of one child who committed suicide after online information was used against him by a school bully observed, "Technology was being utilized as weapons far more effective and reaching [than] the simple ones we had as kids."[201]

In a 2016 article in *Wired* entitled "Why ISIS Is Winning the Social Media War," author Brendan I. Koerner states that "Today the Islamic State is as much a media conglomerate as a fighting force....Never before in history have terrorists had such easy access to the minds and eyeballs of millions." The organization uses online communications platforms such as Twitter, Facebook, Telegram, Surespot, and JustePaste.it on a regular basis to not only share information about their activities but to recruit new members. They have built a successful multiplatform social media campaign that has resulted in 30,000 new recruits from multiple countries. Koerner further notes that unlike other terrorist groups in the past, "the organization does not make this happen in the shadows; it does so openly in the West's most beloved precincts of the Internet, co-opting the digital services that have become woven into our daily lives."[202] ISIS has been very effective at using these services internationally to spread a message of a waiting utopia for those who join their ranks, building an organization with media departments to rival many in the West. They have also crowdsourced messengers and followers, others willing and ready to spread their propaganda on social media, distributing their efforts across the globe and making the group's online presence all but impossible to permanently disrupt. The gunman who used an assault rifle to end many lives in Orlando, Florida, on June 12, 2016, is a recent example of successful online recruitment by ISIS. As one Jordanian official noted, the ubiquitous nature of technology is resulting in the inability to stop this effort by ISIS:

> *Even if I shut down every mosque, every person who supported ISIS in Jordan, there would still be YouTube videos recruiting young men with gun fights that look like they came out of a Hollywood movie. There would still be Twitter where men tweet about how they are living in paradise with three wives and a house, and there would still be WhatsApp and Telegram and every other network for them to communicate personally with whoever they want.*[203]

3.3 Job Losses

As a result of many engineering improvements, entry-level jobs have been lost, with a new term being coined for this effect – "technological unemployment." Automation continues to have the effect of reducing the number of low skill jobs available. In a speech by Andy Haldane, Chief Economist of the Bank of England, he notes that "technology has made it easier and cheaper than ever before to substitute labor for capital, man for machine." Haldane predicts the loss of up to 80 million jobs in the US and 15 million jobs in the UK due to further automation in the future, both in the blue collar and white collar industries.[204]

Nicholas Carr observes in *The Glass Cage* that while the Internet and modern computer technology have opened new doors in terms of opportunities for people to join the global marketplace, many technical skill jobs created as a result may be out of reach for those who have been put out of work due to automation. And it is not only blue collar jobs that face this danger. As intelligent systems begin to take over white collar tasks, many more people may be pushed into lower skill jobs, part time work, or unemployment. Any growth in employment will be seen at the high end and in certain service jobs at the low end, where automation is almost impossible, resulting in a "hollowing out" of middle class jobs. A World Economic Forum report warns that administrative and white collar jobs are at highest risk due to coming advances in artificial intelligence, robotics, and biotechnology as part of a "fourth industrial revolution."[205] As a result of this revolution, it is predicted that we may have five million fewer jobs available globally, bringing mass unemployment and social instability.[206] Nobel Prize winning economist Paul Krugman notes:

Smart machines may make higher GDP possible, but also reduce the demand for people – including smart people. So we could be looking at a society that grows ever richer, but in which all the gains in wealth accrue to whoever owns the robots. [207]

Assembly lines for many products are now becoming increasingly automated, pushing people out of factories in a bid for higher output and greater profit. Chinese companies such as Foxconn, builder of Apple's iPhone, are working on creating more robotic assembly lines with the explicit goal of taking more expensive human labor out of the equation. Job losses are also being driven by the newly engineered web and mobile technologies we have all embraced with gusto. As a current example, Uber, Lyft and other vehicle services are directly affecting taxi drivers, bringing new drivers into service and displacing those who have chosen taxi driving as their profession. Airbnb, HomeAway, FlipKey and other peer-to-peer home rental services are reducing the need for hotels, cutting out all of the jobs associated with their operation – housekeepers, customer service, booking agents, service people, etc. Online legal services such as LegalZoom and Avvu will affect the number of lawyers required to perform many standard legal operations.

As automation grows, the future job market will change dramatically. As the World Economic Forum notes, "around 65% of children starting primary school today will end up working in jobs that don't yet exist." [208] It is worth noting that currently nine million Americans are looking for work, yet 5.4 million job openings are unfilled. [209] Should there be a form of corporate responsibility for companies who engineer technology that results in job losses to ensure an adequate number of people are retrained to take advantage of other opportunities?

3.4 The Wealth/Income Gap and Limited Reach

There are widening wealth and income gaps in countries around the world as societies have embraced technology. In a recent examination of income inequality, the International Monetary Fund (IMF) notes that we are facing issues more serious than many may be aware of:

Widening income inequality is the defining challenge of our time. In advanced economies, the gap between the rich and poor is at its highest level in decades. Inequality trends have been more mixed in emerging

> *markets and developing countries (EMDCs), with some countries experiencing declining inequality, but pervasive inequities in access to education, health care, and finance remain.*[210]

The IMF further notes that income inequality in developing countries has been increasing since 1990. The top 10 percent of wage earners in advanced economies now earn close to nine times as much as the bottom 10 percent. The top one percent of people represents 10 percent of the total income in advanced economies. The wealth gap is also extraordinary. The IMF estimates that almost 50 percent of the wealth in the world is held by just 1 percent of the population. In both the US and China one third of all wealth is held by the top one percent. The percentage of people living in poverty has grown in advanced countries since the 1990s, and in most countries, the growth in wealth of the 1 percent has come at the expense of the bottom 90 percent of people.[211]

But while the gap in income and wealth is clear, how is engineering playing a role in widening it? According to MIT management professor Erik Brynjolfsson, author of *The Second Machine Age,* "My reading of the data is that technology is the main driver of the recent increases in inequality. It's the biggest factor."[212] Stephen Hawking agrees, noting,

> *If machines produce everything we need, the outcome will depend on how things are distributed. Everyone can enjoy a life of luxurious leisure if the machine-produced wealth is shared, or most people can end up miserably poor if the machine-owners successfully lobby against wealth redistribution. So far, the trend seems to be toward the second option, with technology driving ever-increasing inequality.*[213]

Brynjolfsson also warns that rapid growth in technology is making jobs redundant at a pace greater than it is replacing them.

> *Productivity is at record levels, innovation has never been faster, and yet at the same time, we have a falling median income and we have fewer jobs. People are falling behind because technology is advancing so fast and our skills and organizations aren't keeping up.*[214]

Can engineers help find a way to help narrow this gap instead of widening it? Or will the automated future where jobs are scarce for the lower echelon of society result in a much greater gap, where only the developers and operators of the automated systems are the ones getting richer? As engineering has led to greater wealth creation, the ability of the rich to create both physical and financial havens has led to inequality and social discord. The divide between the rich and the poor is widening as technologies and the utilization of wealth itself increase the ability of the rich to build greater resources. As newer technologies become more complex and also more expensive, there is a growing divide not just inside communities but both within and between developed and developing countries. How will those in developing countries afford the technology that is considered a required part of a normal existence in the developed world, and what might the consequences be when the gap widens so far that those who have not target those who have?

3.5 Health

Technology is changing human habits in a way unseen in our history, as the best of our inventions are supporting the worst of our habits. A recent article in the *Washington Post* notes that Internet and gaming addiction amongst young people is no longer just an issue in countries like South Korea and China, but is also now a significant factor in the US.[215] Overindulgence in the worlds of gaming and Internet usage is causing some to drop out of school, disconnect from friends and family, and withdraw from their lives. The dopamine boost in their brains from the positive experiences they have online becomes an addiction, and these individuals show symptoms very similar to other forms of addiction. These include an inability to stop an activity, irritability, depression, and a decline in physical health due to a lack of proper nutrition. In China, Japan and South Korea these addictions are officially recognized and treated, while in the US there is still no official classification. As the developed world begins to experiment with new forms of virtual reality technology, it is expected that an addiction to existence in a virtual world may also occur.

Fast and heavily processed food, with its excess sugars, fats and preservatives, coupled with optimized food storage, have made food available anywhere, anytime. Factories churn through mountains of ingredients, blended, cooked and packaged to sustain people by the millions. Taking advantage of economies of scale, this food is extremely cheap and is engineered to be delicious, calorie-rich and easily accessible. Although fast food has become an important part of our lives, processed to conveniently satiate us at a low cost, it has also brought with it many side effects that have had a negative impact on our health. The engineering breakthroughs that may appear positive have also resulted in a health crisis throughout the developed world.

Overconsumption has created an increase of type 2 diabetes, obesity, stroke, and heart attacks,[216] as most fast foods are high in calories from unhealthy fats and sugar in addition to larger portion sizes. The increased popularity of "engineered food" has resulted in more than doubled obesity rates among adults in the US in the past 35 years. The average American is now 24 lbs. heavier today than in 1960. Similarly, childhood obesity has tripled since 1980.[217]

Unacceptable levels of antibiotics have been found in the meats used in many fast food restaurants, adding to the health crisis we already face from the rise of antibiotic-resistant bacteria. Animal growth hormones are found in the meat used in fast food and other restaurants as a byproduct of our need to grow animals larger and faster.[218] A recent study from the George Washington University notes that those who consumed fast food had high levels of two phthalates, di(2-ethylhexyl) phthalate (DEHP) and diisononyl phthalate (DiNP). These are chemicals typically used to add flexibility to plastics, and have been suspected of deleterious health effects including male reproductive system development problems, birth defects, childhood behavioral problems and chronic childhood illnesses.[219] Rates of obesity and diabetes are climbing overall, but particularly in low-income communities, with greater intake of fast and packaged food due to its low cost, availability, addictive flavoring, and shelf life. Andrea Freeman of the California Law Review labels fast food as "oppression through poor nutrition," as our success in creating cheap but unhealthy food unfairly impacts poor and minority communities.[220]

Our successful export of high-flavor, calorie-dense fast food is already affecting other countries where traditional diets are giving way to modern convenience. Pan, Malik and Hu, in their paper *Exporting Diabetes to Asia: The Impact of Western-style Fast Food,* examined this phenomenon. They observed the tremendous growth of fast food companies throughout Asia and the resulting effects on the health of Asian populations. They found that greater consumption of fats, salt and sugar in larger portion sizes along with lax public policies are driving up traditionally low rates of diabetes, obesity and cardiovascular heart disease.

> *If measures are not taken to abate these trends, the twin epidemics of obesity and diabetes in Asian countries could have the potential to overwhelm fragile health care systems and counteract remarkable achievements in economic growth....While globalization has undoubtedly resulted in some beneficial changes to society, its*

unintended consequences are driving the worldwide obesity epidemic.[221]

Certain materials have been adopted with gusto in engineered products due to their favorable properties, but have later been found to have a deleterious effect on health. Asbestos is a great example of this. Once considered a good choice as a building material due to its many positive properties, asbestos fibers were later found to become easily airborne when disturbed, causing life-threatening diseases when inhaled. These include lung cancer, asbestosis, and mesothelioma, a cancer found in the lining of the thoracic and abdominal cavities.[222] DDT, or dichlorodiphenyltrichloroethane, used as an agricultural insecticide for many years, is another example. It was banned after it was found to cause breast and other cancers, male infertility, miscarriages, low birth weights, developmental delays, nervous system damage, and liver damage.[223] There are in fact so many materials that are now known or suspected to be carcinogenic that the National Institutes of Health (NIH) publishes a report long enough that a simple list of their names fills five pages. Included on the "known to be carcinogen" list are many standard items found in engineering such as lead, mineral oils, wood dust, nickel compounds, and radon.[224]

We are also facing a crisis of inequality from our engineering successes in the health industry. As we add greater cost and technological complexity to our medical devices, lifesaving operations are only available to those who can afford them. For those who do not have the funds to cover complex technology-driven surgeries or expensive scans, their lives may be cut short, even when we possess the ability to prolong them. As a result, the increasingly unaffordable cost of our modern high-technology medicine practice is having life and death effects on some sections of society.

The cost of medical technology in our country is steadily growing. Purchasing a hospital bed can now cost up to $40,000,[225] patient monitors cost up to $35,000,[226] a state of the art MRI machine can cost up to $3 million,[227] a 320-slice CT scanner up to $2.5 million, and an x-ray machine $125,000.[228] Even a 42" television, at around $400 for a consumer level model, is almost $4,000 for a medical grade version (costs are often inflated to cover liability risks).[229] It is little surprise that with expensive equipment costs, hospitals are forced to pass along these expenditures to stay in business. The cost of an average 3-day hospital stay has grown to $30,000.[230] This is a combined policy, business and engineering problem that must be addressed simultaneously.

3.6 Finance

The days of human-executed trades on the floor of the New York Stock Exchange (NYSE) are mostly a thing of the past, as Wall Street has brought in technology to take their place. These technologies now handle between 50 and 75 percent of stock trades performed every day.[231]

As a consequence of increased computational power and the development of sophisticated new analysis techniques, high frequency traders are now using technology to leverage latency delays in the milliseconds that computerized trading systems take to communicate trades. This allows them to gain pennies or fractions of pennies on each stock share on big trade volumes, thereby making up to an estimated $3 billion in profit each year. A suggested method to correct this fundamental flaw in the setup of the US stock exchange system is to introduce a "speed bump" – forcing communications signals to traverse 38 physical miles of cable in order to slow them by a few milliseconds.[232] These high frequency traders are also contributing to the overall volatility of the market, using complex algorithms, automation, and moving huge stock volumes in very short periods, and this can have affect everyone in the market, including individual investors. In the event that a company's trading system fails and executes incorrect trades, the automation of the rest of the market can lead to a chain reaction or "flash crash" that can have significant negative financial effects.

> *This tangle of systems is so complicated that its behavior often appears irrational. Even financial experts have a hard time understanding it. A crucial reason is that the most pernicious problems – latency arbitrage, flash crashes – are not, first and foremost, financial problems. They are computer science problems.*[233]

Gambling, once restricted primarily to casinos with dealers, cards, and manual mechanical slot machines, has now gone both high-tech and online. While there is a moral discussion to be had around gambling, there is no questioning the fact that modern electronic slot machines provide the ability to very efficiently and effectively empty the pockets of those who have an addiction.

> *Slot machines are nowadays sophisticated computerized devices engineered to produce continuous and repeat betting, and programmed by high-tech experts to encourage gamblers to make multiple bets*

simultaneously by tapping buttons on the console as fast as their fingers can fly.[234]

Casinos also use large amounts of data collected via loyalty cards to leverage analysis and prediction techniques, ensuring gamblers keep coming back for more. The more that technology improves access to gambling in more forms, the more it victimizes those who least can afford it. In a Consumer Federation of America poll, it was found that 2 out of 5 people earning less than $25,000 a year believe playing the lottery is the most practical way to build wealth![235] While the US allows legal online gambling only in Nevada, Delaware, and New Jersey, the global online gambling market is valued at $37 billion per year across 85 countries, with growth predicted to be 11% per year for at least the next four years.[236]

The global reach of electronic funds transfer technologies has enabled teams of international scammers to target millions of people, convincing them through social engineering practices to transfer large sums of money to criminal organizations around the world. The use of the private electronic money transfer and communication system known as SWIFT (Society for Worldwide Interbank Financial Telecommunication) revolutionized money transfer internationally among its 11,000 member institutions in more than 20 countries. The network carries 27.5 million messages a day, but much like any technology, it too has its flaws. In 2016 the Central Bank of Bangladesh reported that it had lost $81 million to thieves utilizing the SWIFT system via malware installed on bank computers. This followed a $12 million loss at the Banco del Austro in Ecuador in 2015 and many others from banks in Southeast Asia.[237]

3.7 Human Intelligence

Technology was in many ways intended to unlock human potential by freeing us from much of our labor time. But is it improving or harming our minds? Engineering has created the ability for human beings to access information 24/7, from almost anywhere. This should make us smarter, but the unintended consequence is that we're actually becoming less intelligent as a result. Phrases such as "TL:DR", shorthand for "too long: didn't read," represent the negative effect of consuming small snippets of information on a regular basis rather than developing a true understanding of content. A cognitive expert in *The Glass Cage* observes that on one hand we have greatly developed the cognitive skill of skimming and identifying useful information very quickly, but it has come at the cost of our ability to focus, dig deeper and develop a full understanding of the subject matter.[238]

How is intelligence defined? Many voices have provided complex answers, but one definition is that we can learn, pose, and revise problems, all of which have some complexity.[239] It is with this definition that we examine the modern cash register at a fast food restaurant as an example, with its ability to provide simple answers to minimally complex problems without the requirement of human problem-solving. Well-engineered systems have led to a simplistic end-user experience such that more of the population can access and use an order entry system with very little knowledge of what they are doing. While the growth of smart cash registers is allowing for more jobs in the entry-level low skills sector, it is also encouraging lower requirements for basic knowledge in society. If a machine can do the math, why does one need to develop math skills to calculate the change on an order? Better yet, if the cash register is replaced with the latest in phone-based payment systems, nothing is required of the cashier other than to press the OK button during a transaction. With the proliferation of various "simplified" devices designed for use by the least skilled of our workforce, we are robbing them of the opportunity to learn and develop their intelligence. The growth of technology with simple interfaces may also be making us "simpler," instead of forcing us to learn new tools and techniques in order to accomplish these tasks.

As another example in *The Glass Cage*, author Nicholas Carr considers the long-term effect of GPS usage, which results in a lack of path-building exercise in the brain, perhaps causing an overall decrease in memory and even contributing to the early onset of dementia. He also notes the concern surrounding increased automation and the unexpected effect of a breakdown of the human skills required when disaster strikes, such as having to take control when piloting a heavily automated aircraft.[240]

The wide range of unintended consequences of our often-unrestrained search for greater technological achievement will in many ways define the future of engineering. While engineers have developed tremendous capabilities to create products and systems at a scale and speed once unimaginable, we have been careless with the fruits of their labor. When goods became cheaper and ubiquitous, those who could afford them engaged in overconsumption without constraint. As a result, our natural environment has been compromised, with the impacts of our excess even at the furthest reaches of our planet. While we have adopted the Internet at a rate greater than any other technology in history, our nature as human beings has led us to misuse this abundant, uncontrolled resource. The march of technology has led to job losses, our health and intelligence has become compromised by our modern way of life, and we have a growing gap between the rich and the poor. How will society perceive engineers in future if these unintended consequences are not mitigated? In the words of Lee Iacocca, "We are continually faced by great opportunities brilliantly disguised as insoluble problems."[241] Clearly, we need to consider unintended consequences as we teach engineering design in our undergraduate curricula. It should nevertheless be noted that the importance of unintended consequences depends upon where you are on the economic ladder. For advanced countries they represent significant negativity, whereas a less advanced nation turns a jaundiced eye. Our students should know that as humanity faces the challenges from climate change, waste, pollution, etc., engineers will be called upon to not only rectify past mistakes, but also make forward-looking decisions to avoid creating unmanageable unintended consequences.

References

[171] Helle, K., "Rural Electrification", Digital Image, Flickr, December 10, 2008, accessed at https://www.flickr.com/photos/deepeco/15841489706

[172] Jessica Lea/DFID, "Drone surveillance helps search and rescue in Nepal", Digital Image, Flickr, April 29, 2015, accessed at https://www.flickr.com/photos/dfid/16691214064

[173] Attributed to Albert Einstein, Theoretical Physicist, March 14, 1879 – April 18, 1955.

[174] Qui, J., "Man-made pollutants found in Earth's deepest ocean trenches", *Nature*, June 20, 2016, accessed at http://www.nature.com/news/man-made-pollutants-found-in-earth-s-deepest-ocean-trenches-1.20118

[175] Bowie, L., "Water from a fountain? Not in Baltimore city schools", *The Baltimore Sun*, April 9, 2016, accessed at http://www.baltimoresun.com/news/maryland/baltimore-city/bs-md-ci-lead-in-water-20160409-story.html

[176] Young, A., Nichols, M., "Beyond Flint: Excessive lead levels found in almost 2,000 water systems across all 50 states", *USA Today*, 2016, accessed at http://www.usatoday.com/story/news/2016/03/11/nearly-2000-water-systems-fail-lead-tests/81220466/

[177] Biello, D., "Fertilizer Runoff Overwhelms Streams and Rivers – Creating Vast "Dead Zones"", *Scientific American*, March 14, 2008, accessed at http://www.scientificamerican.com/article/fertilizer-runoff-overwhelms-streams/

[178] "Maryland's New Lawn Fertilizer Law Takes Effect October 1", Maryland Department of Agriculture, September 26, 2013, accessed at http://news.maryland.gov/mda/press-release/2013/09/26/marylands-new-lawn-fertilizer-law-takes-effect-october-1/

[179] Brown, P.L., "The Flint of California", Politico, May 25, 2016, accessed at http://www.politico.com/agenda/story/2016/05/is-clean-drinking-water-a-right-000129

[180] Thompson, D., "2.6 Trillion Pounds of Garbage: Where Does the World's Trash Go?", *The Atlantic*, accessed at http://www.theatlantic.com/business/archive/2012/06/26-trillion-pounds-of-garbage-where-does-the-worlds-trash-go/258234/

[181] "Leachate", Wikipedia, accessed at https://en.wikipedia.org/wiki/Leachate

[182] Koerth-Baker, M., "How Do You Put Out A Subterranean Fire Beneath A Mountain Of Trash?", FiveThirtyEight, May 10, 2016, accessed at http://fivethirtyeight.com/features/how-do-you-put-out-a-subterranean-fire-in-a-mountain-of-trash/

[183] Lakshmi, R., "A burning mountain of trash in Mumbai fuels middle-class outcry", *The Washington Post*, April 15, 2016, accessed at https://www.washingtonpost.com/world/asia_pacific/a-burning-mountain-of-trash-in-mumbai-fuels-middle-class-outcry/2016/04/11/28336511-33b4-437d-8f4d-f53dba9fdf0e_story.html

[184] Ibid.

---

[185] Parker, L., "Ocean Trash: 5.25 Trillion Pieces and Counting, but Big Questions Remain", *National Geographic*, January 11, 2015, accessed at http://news.nationalgeographic.com/news/2015/01/150109-oceans-plastic-sea-trash-science-marine-debris/

[186] Jambeck, J.R., Andrady, A., Geyer, R., Narayan, R., Perryman, M., Siegler, T., Wilcox, C., Lavender Law, K. ,"Plastic waste inputs from land into the ocean", *Science,* Issue 347, pp. 768-771, 2015.

[187] "Congress Oks bill banning plastic microbeads in skin care products", CBS News, December 18, 2015, accessed at http://www.cbsnews.com/news/congress-oks-bill-banning-plastic-microbeads-in-skin-care-products/

[188] Lewis, D., "Five Things to Know About Congress' Vote to Ban Microbeads", *The Smithsonian*, December 10, 2015, accessed at http://www.smithsonianmag.com/smart-news/five-things-know-about-congress-vote-ban-microbeads-180957501/

[189] "Great Pacific Garbage Patch", Wikipedia, accessed at https://en.wikipedia.org/wiki/Great_Pacific_garbage_patch

[190] Schwartz J., "Study Finds Rising Levels of Plastics in Oceans", *The New York Times*, February 12, 2015, accessed at http://www.nytimes.com/2015/02/13/science/earth/plastic-ocean-waste-levels-going-up-study-says.html

[191] "The Future of Silicon Valley", CNBC, accessed at http://video.cnbc.com/gallery/?video=3000242076

[192] "Policy Brief on E-waste What, Why and How", United Nations Envrionment Programme, May 13, 2013, accessed at http://www.unep.org/ietc/Portals/136/Other%20documents/PolicyBriefs/13052013_E-Waste%20Policy%20brief.pdf

[193] Campbell, K., "Where does America's e-waste end up? GPS tracker tells all", PBS Newshour, May 10, 2016, accessed at http://www.pbs.org/newshour/updates/america-e-waste-gps-tracker-tells-all-earthfix/

[194] "Induced Earthquakes", United States Geological Survey, April 6, 2016, accessed at http://earthquake.usgs.gov/research/induced/

[195] Jackson, R., et. al., "Increased stray gas abundance in a subset of drinking water wells near Marcellus shale gas extraction", *Proceedings of the National Academy of Sciences of the United States of America*, Volume 110, No. 28, July 9, 2013.

[196] Pecht, M.G., Kaczmarek, R., Song, X., Hazelwood, D., Kavetsky, R., & Anand, D., *Rare Earth Materials: Insights and Concerns*, CALCE EPSC Press, p. 75, 2011.

[197] "Household air pollution and health", World Health Organization, February 2016, http://www.who.int/mediacentre/factsheets/fs292/en/

[198] "Milky Way now hidden from a third of humanity", National Oceanic and Atmospheric Administration, June 10, 2016, accessed at http://www.noaa.gov/stories/milky-way-now-hidden-third-humanity

---

[199] "IC3 2014 Internet Crime Report", United States Federal Bureau of Investigation, accessed at https://www.fbi.gov/file-repository/2014_ic3report.pdf/view

[200] Timberg, C., "A flaw in the design", *The Washington Post*, May 30, 2015, accessed at http://www.washingtonpost.com/sf/business/2015/05/30/net-of-insecurity-part-1/

[201] "The Top Six Unforgettable Cyberbullying Cases Ever", Nobullying.com, June 23, 2016, accessed at https://nobullying.com/six-unforgettable-cyber-bullying-cases/

[202] Koerner, B.I., "Why ISIS is Winning the Social Media War", *Wired*, April, 2016, accessed at https://www.wired.com/2016/03/isis-winning-social-media-war-heres-beat/

[203] Frenkel, S., "Everything You Ever Wanted to Know about How ISIS Uses the Internet", BuzzFeed, May 12, 2016, accessed at https://www.buzzfeed.com/sheerafrenkel/everything-you-ever-wanted-to-know-about-how-isis-uses-the-i

[204] "Labour's Share - speech by Andy Haldane", Bank of England, November 12, 2015, accessed at http://www.bankofengland.co.uk/publications/Pages/speeches/2015/864.aspx

[205] Carr, N., *The Glass Cage: Automation and Us*, W. W. Norton & Company, 2014.

[206] Kottasova, I., "Technology could kill 5 million jobs by 2020", CNN Money, Jnauary 18, 2016, accessed at http://money.cnn.com/2016/01/18/news/economy/job-losses-technology-five-million/

[207] Krugman, P., "Is Growth Over?", *The New York Times*, December 26, 2012, accessed at http://krugman.blogs.nytimes.com/2012/12/26/is-growth-over/

[208] Kottasova, I., "Technology could kill 5 million jobs by 2020", CNN Money, January 18, 2016, accessed at http://money.cnn.com/2016/01/18/news/economy/job-losses-technology-five-million/

[209] Andrews, L.M., "Robots Don't Mean the End of Human Labor", *The Wall Street Journal*, August 23, 2015, accessed at http://www.wsj.com/articles/robots-dont-mean-the-end-of-human-labor-1440367275

[210] Dabla-Norris, E., et al., "Causes and Consequences of Income Inequality: A Global Perspective", International Monetary Fund, June, 2015, accessed at https://www.imf.org/external/pubs/ft/sdn/2015/sdn1513.pdf

[211] Ibid.

[212] Rotman, D., "Technology and Inequality", *MIT Technology Review*, October 21, 2014, accessed at https://www.technologyreview.com/s/531726/technology-and-inequality/

[213] Hawking S., "Science AMA Series: Stephen Hawking AMA Answers!", Reddit, 2015, https://www.reddit.com/user/Prof-Stephen-Hawking

[214] Rotman, D., "How Technology Is Destroying Jobs", *MIT Technology Review*, June 12, 2013, accessed at

https://www.technologyreview.com/s/515926/how-technology-is-destroying-jobs/

[215] Tsukayama, H., "This dark side of the Internet is costing young people their jobs and social lives", *The Washington Post*, May 20, 2016.

[216] Wiess, T.C., "Associated Health Risks of Eating Fast Foods", Disabled World, March 31, 2016, accessed at http://www.disabled-world.com/fitness/fast-food.php

[217] "Obesity Rates & Trends Overview", Trust for America's Health and The Robert Wood Johnson Foundation, 2016, accessed at http://stateofobesity.org/obesity-rates-trends-overview/

[218] Tinker, B., "Restaurant report card grades on antibiotics in meat supply", Cable News Network, September 16, 2015, accessed at http://www.cnn.com/2015/09/15/health/fast-food-meat-antibiotics-grades/

[219] "Fast Food may come with a side of Pthalate Chemicals", CBS News, April 13, 2016, accessed at http://www.cbsnews.com/news/fast-food-may-come-with-a-side-of-phthalate-chemicals/

[220] Freeman, A., "Fast Food: Oppression through Poor Nutrition", *California Law Review,* Volume 95, Issue 6, Article 8, December 12, 2007.

[221] Pan, A., et al., "Exporting Diabetes to Asia: The Impact of Western-Style Fast Food", *Circulation*. Volume 126, Issue 2, July 10, 2012, pp. 163–165.

[222] "Asbestos", How Products Are Made, 2016, accessed at http://www.madehow.com/Volume-4/Asbestos.html

[223] "Eskenazi, B. et. al, "The Pine River Statement: Human Health Consequences of DDT Use", *Environ Health Perspect.,* Volume 117, Issue 9, pp. 1359–1367, May 4, 2009.

[224] Substances Listed in the Thirteenth Report on Carcinogens", National Institutes of Health, 2016, accessed at http://ntp.niehs.nih.gov/ntp/roc/content/listed_substances_508.pdf

[225] Rubenfire. A., "Hospitals paying more for electric beds", Modern Healthcare, April 27, 2015, accessed at http://www.modernhealthcare.com/article/20150427/NEWS/150429935

[226] Blankehorn, D., "How Philips maintains monitor price points", ZDNet, October 27, 2010, accessed at http://www.zdnet.com/article/how-philips-maintains-monitor-price-points/

[227] Glover, L., "Why Does an MRI Cost So Darn Much?", *Time*, July 16, 2014, accessed at http://time.com/money/2995166/why-does-mri-cost-so-much/

[228] Mustapha, T., "Your Guide to Medical Imaging Equipment", Block Imaging, July 2, 2014, accessed at http://info.blockimaging.com/ge-digital-radiography-system-price-cost-guide

[229] "Generation II 42" RoomMate Hospital-Grade LCD IPTV", 4MD Medical, 2016, accessed at http://www.4mdmedical.com/generation-ii-42-roommate-hospital-grade-lcd-iptv.html

[230] "Protection from high medical costs", Healthcare.gov, accessed at https://www.healthcare.gov/why-coverage-is-important/protection-from-high-medical-costs/

[231] Arthur, G., "Why do they still have floor traders at the NYSE?", NPR Marketplace, July 4, 2014, accessed at http://www.marketplace.org/2014/07/04/business/ive-always-wondered/why-do-they-still-have-floor-traders-nyse

[232] Lam, B., "A New, Slower Stock Exchange", *The Atlantic*, June 20, 2016, accessed at http://www.theatlantic.com/business/archive/2016/06/iex-approved/487898/

[233] Groskoph, C., "The modern stock market is a badly designed computer system", Quartz, June 15, 2016, accessed at http://qz.com/662009/the-sec-tried-to-fix-a-finance-problem-and-created-a-computer-science-problem-instead/

[234] Whitehead, B.D., "Gaming the Poor", *The New York Times*, June 21, 2014, accessed at http://opinionator.blogs.nytimes.com/2014/06/21/gaming-the-poor/

[235] Spurlock, M., "Inside Man: Double or Nothing", Video, Cable News Network, aired June 3, 2016.

[236] "Global Online Gambling Market Growing At 11-Percent Clip", Card Player, February 26, 2016, accessed at http://www.cardplayer.com/poker-news/20036-global-online-gambling-market-growing-at-11-percent-clip

[237] Zetter, K., "That Insane, $81M Bangladesh Bank Heist? Here's What We Know", May 17, 2016, accessed at https://www.wired.com/2016/05/insane-81m-bangladesh-bank-heist-heres-know/

[238] Carr, N., *The Glass Cage: Automation and Us*, W. W. Norton & Company, 2014.

[239] Gardner, H., Perkins, D., Sternberg, R., "Theories of Intelligence", University of Oregon, accessed at http://otec.uoregon.edu/intelligence.htm

[240] Carr, N., *The Glass Cage: Automation and Us*, W. W. Norton & Company, 2014.

[241] Widely attributed to Lee Iacocca.

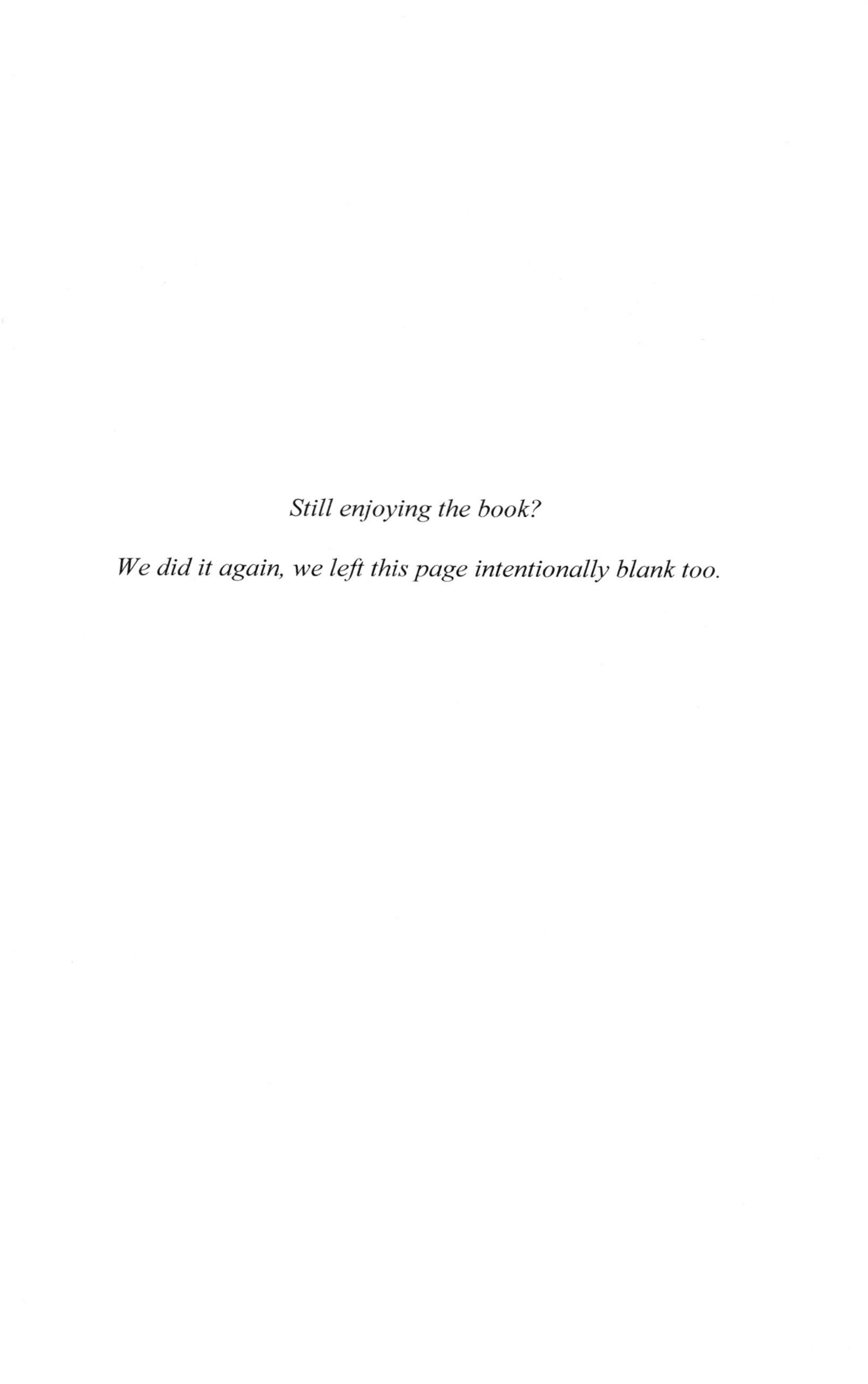

Still enjoying the book?

We did it again, we left this page intentionally blank too.

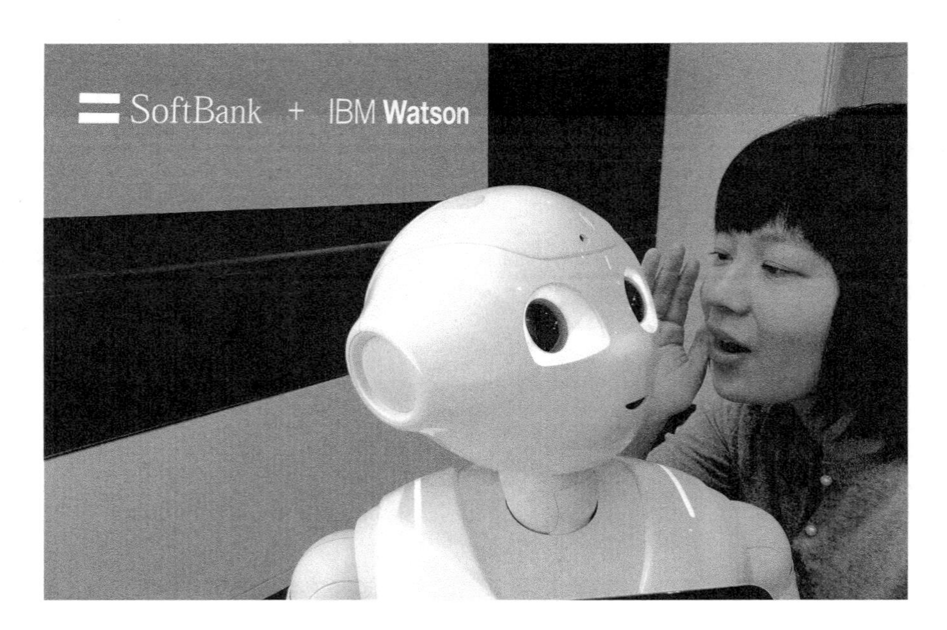

An IBM Tokyo researcher interacts with SoftBank's Pepper robot, Powered by Watson AI software [242]

A NASA crew in the melting Arctic sea studying climate impacts [243]

Chapter 4

Game Changers

Prediction is very difficult, especially if it's about the future.[244]

Whereas we discussed catalysts for social change that describe our current state of affairs in Chapter 2, in this chapter we examine the future of various sectors and associated disruptive technologies. We also consider future possibilities in terms of the social impact that may result from engineering breakthroughs in these areas. Even with the tremendous social change engineering has created in recent decades, we believe there are areas where engineering will create even greater social impact, both with current technologies and those that we are now developing. As examples we discuss artificial intelligence, transportation, communications, medicine, big data and crowdsourcing, climate change, and robotics.

4.1 Artificial Intelligence

With the increasing introduction of autonomous systems, artificial intelligence (AI) will become an integral part of our daily lives. While the true meaning of AI is often considered total independence from human intervention, we are already witnessing shades of AI in its infancy. When fully developed and deployed, the impact of highly intelligent AI systems on society will be transformative, and some believe they may even threaten our very existence. Ray Kurzweil asks, "Can an intelligence create another intelligence more intelligent than itself?"[245]

We currently have digital assistants in Apple's Siri, Microsoft's Cortana, Amazon's Alexa, and Google Now performing close to real-time voice recognition and semi-intelligent tasks. The Google search engine uses predictive analytics to know what you are looking for by just typing one or two keywords. Netflix suggests movies to watch based on algorithms that use data from your previous viewing habits. These are simplistic forms of AI driven by entire buildings full of computational servers via a simplified consumer-friendly interface. At Apple's 2016 Worldwide Developers Conference, it was noted that the new facial recognition process used to simply match facial features and group personal photos of different people requires 11 billion computations per photo in their data centers![246]

In 2011 IBM's intelligent computer system "Watson" defeated a champion of the game show *Jeopardy* and set a landmark for AI. Since Watson analyzes unstructured data, and noting that nearly 80% of all digitized information today is unstructured, Watson and systems like it are good predictors of what the future holds in the area of AI.[247] In 2016 Deepmind, an AI system currently being developed by Google, was able to beat a human player at the game AlphaGo for the first time ever. Due to the nature of the game, this is considered a much more difficult feat than beating a human at chess, a previous achievement of another AI system.[248] Deepmind utilizes a process of machine learning called deep learning, where a neural network is built from vast quantities of data.

While AI is most often portrayed as being best suited for commerce or general technology purposes, the Chinese company Baidu has adapted its newest deep learning AI system for use in a product for social good. DuLight for the blind, a headset with a camera and earpiece, uses the Cloud for AI-based near-realtime image recognition and conveys it back to the user in audio feedback. The system's massive data centers are being constantly fed millions of images to be able to learn to recognize objects the user may encounter. As with most deep learning systems, the more data that is fed into the system over time, the more accurate the system will become.[249]

It is projected that in future, AI will find applications in almost every human endeavor. In education, virtual and e-learning will become more pronounced where educators will adapt their teaching methodologies to best meet the learning ability of the individual using "intelligent" assistive technologies. Traditional teachers will be considered mentors rather than a source for knowledge. Moving further ahead, virtual institutions may then be established that would enable the individual to receive a world-class education regardless of their geographical location and socioeconomic background. AI systems would intelligently build updated and personalized course materials based on individual learner profiles, automatically evaluating and assisting students using alternative pedagogical approaches in the event a student is experiencing learning difficulties.

In future medical systems, AI will be more widely implemented, monitoring patient health metrics and using DNA analysis to predict possible future medical conditions. AI will be widely used by medical personnel to intelligently evaluate medical tests and reports based on vast stores of historical data, thus improving treatment efficiency.[250] With access to a wide range of medical data in real time, AI-driven systems will become key diagnostic and treatment tools for medical personnel in their daily work. In many ways AI will also enable people to track and predict their own health without the need for visits to a doctor's office, and as a result, the average human life span in the developed world can be expected to increase.

While vast stores of interconnected information in combination with intelligent systems will certainly improve the medical field overall, some believe the future implementation of AI in healthcare may lead to closure of small clinics and healthcare centers. Instead, medical consultations will be conducted with "virtual doctors" or smart systems through mobile apps and other technology tools. Given predicted future availability of intelligent online medical technologies, people may not seek medical care unless they are very sick or require surgery, which then may be performed by an automated robotic surgery device.[251] As in other industries where AI is being applied, the future may see a reduced need for medical personnel and jobs may be lost.

It is fair to say that the use of AI will be ubiquitous. We will see "intelligent" factories running autonomously, sophisticated platforms will evolve in financial systems, and there will be enhanced control for such activities as high frequency trading (HFT).[252] Finally and importantly, computer network security will be implemented using AI's predictive capabilities to prevent abrupt system failures and sabotage. Public security will be enhanced as a combination of widespread video cameras, a vast network of IOT sensors, facial recognition, and AI's deep learning will enable comprehensive surveillance in developed countries, challenging policymakers and society.

While there are many prominent voices in our society raising red flags surrounding the future development and possible dangers of advanced AI,[253] Ray Kurzweil, the widely recognized inventor, futurist, and Director of Research for Google, argues, "As AI continues to get smarter, its use will only grow. Virtually everyone's mental capabilities will be enhanced by it within a decade".[254] He notes that a child in Africa using a smartphone has access to more information than the President of the United States two decades ago, and that as AI assists us in addressing the world's grand challenges, we should be vigilant instead of fearful. With a growing world population and pressing health–, energy–, and climate–related concerns, we must utilize these systems to enhance our productivity and design sustainable solutions.

However, Steven Hawking and others have expressed serious concerns. Hawking notes that if true AI were to be developed, machines would be capable of rapid prototyping and improvement to the point where "humans, who are limited by slow biological evolution, couldn't compete, and would be superseded."[255] Steve Wozniak notes that "If we build these devices to take care of everything for us, eventually they'll think faster than us and they'll get rid of the slow humans to run companies more efficiently."[256]

An open letter was signed in July 2015 by more than a thousand robotics and AI researchers warning of the dangers of combining AI and warfare.[257] With the words of Kurzweil in mind, that technology is a double-edged sword, and whether one is for or against the development of AI there are a few questions we need to consider:

- Accountability – If a system has a mind of its own and comes to conclusions through reasoning and not based on data it is fed, who is accountable for the outcome?
- Security – AI will exist as programming within a computer system, and thus how vulnerable is the system to hacking and system failure?
- Conflict – What would happen if the world's military were to use intelligent autonomous weapons systems? Would that pave the path for a safer and peaceful world with less war and fewer battlefield casualties, or the polar opposite?
- Human Obsolescence – If an AI system ever comes to a point where it has higher intelligence than humans do, who will then be in command? Will AI machines be designing next generation AI systems, leaving humans behind?

While the notion of autonomous weapons systems brings a lot of hope to the warfighter on the ground for a safer battlefield (where men and women on active duty currently operate under extraordinarily pressing conditions), it also raises concerns for the outcome of self-operating, state-of-the-art lethal weapons with no human component.[258]

In support of AI skeptics, recall the glitch in the computerized trading system of the firm Knight Capital Group in 2012 that led the company to a $440 million loss in just 30 minutes. The root cause was found to be an algorithm gone awry, with erroneous trades in 150 different stocks.[259] A system with the ability to make autonomous decisions, the power to enforce them, and no human in-the-loop is a serious concern.

Yoshua Bengio, a leading deep learning AI researcher, sees the perceived dangers of the growth of AI as overblown, noting that systems are still so far short of intelligence that the field should be referred to as "artificial stupidity." He is less concerned about the future of AI in terms of intelligent computer systems taking over the world and causing harm to humanity, but instead expresses his concern over how people might take advantage of AI. "I'm more worried about the misuse of AI. Things like bad military uses, manipulating people through really smart advertising, also, the social impact, like many people losing their jobs. Society needs to get together and come up with a collective response, and not leave it to the law of the jungle to sort things out."[260]

Perhaps a case in point was Microsoft's recent AI bot, given a young female personality and named Tay, which was then set loose to autonomously communicate with people on messaging apps Twitter, Kik and GroupMe. As Tay had AI abilities, the system learned from its communications and within 24 hours transformed from a flirty teen to a sexist, racist monster. Clearly, Microsoft engineers thought little about the nature of how people interact with machines.[261]

We may be closer than expected to developing exponentially more capable AI systems. As Moore's Law is quickly approaching a physical boundary in transistor size within conventional processors, quantum computing is predicted to provide the capability for extreme computation needs, such as pattern recognition and machine learning for AI applications. Based upon the work of the Canadian company D-Wave Systems, which built an experimental quantum computing system demonstrating a test running at 100 million times the speed of a conventional processor, Google has formed a Quantum Artificial Intelligence Lab with the goal of using superconducting chilled aluminum loops to demonstrate a high quality universal quantum computer in two years from now.[262] And recently, IBM made quantum computing publicly available with its cloud-enabled platform called The IBM Quantum Experience,[263] providing simplified public access to a five qubit machine that utilizes superconducting circuits in a sub-zero refrigerator. The company hopes to encourage experimentation on the platform to discover how to best leverage quantum capabilities.[264]

With all of the concerns surrounding the untethered development of AI, a nonprofit called OpenAI has been created and sponsored by a group including Elon Musk and PayPal co-founder Peter Thiel, with a pledged $1 billion in funds. The organization is described as an AI research company whose goal is to "advance digital intelligence in the way that is most likely to benefit humanity as a whole, unconstrained by a need to generate financial return."[265]

4.2 Transportation

By virtue of improved transportation, our global society is better connected today than ever before, with improved communications, trading and mobility.[266] Technical advances in transportation technology have resulted in a continual decrease in the overall cost of transportation since the 1930's.[267] Improved transportation systems have enhanced cross-border intellectual and cultural exchanges in societies around the world. Many have developed an appreciation for diversity from interaction with different cultures, and this newfound mobility has allowed them to venture far beyond their local communities to gain education and/or career opportunities. Local transportation systems have also lowered barriers between individuals from disparate communities. Today, even those with lower socioeconomic status are able to travel long distances in some countries at low cost.

Future transportation systems include autonomous vehicles in the form of cars, planes, trains, and buses, as well as entirely new forms of transportation, such as Elon Musk's Hyperloop concept. Danielle Muoio examines technologies we may utilize in 2045 and notes that "in the next 30 years, though, we are likely to see more change in transportation technology than we've seen in the last 100 years."[268] This will come at a price, requiring us to accept changes that may include a tremendously disruptive overhaul of our transportation infrastructure. We will also face significant changes in societal patterns as a result, particularly in how, when, and where people work, live, and play. Much of this change will be positive, as in the future our transportation systems may also be designed to best meet the needs of the individual traveler. Physical disabilities will no longer prevent people from travelling, due to evolving assistive technologies, both on their person and integrated into future modes of transportation.[269]

How will our world change in a positive way following significant technological advances in the transportation industry? The number of traffic accidents may be significantly reduced, as driverless cars would remove human-specific errors. Traffic issues would become less of a concern, as sensors and vehicle-to-vehicle communication systems would allow for more efficient routing. Driverless vehicles would encourage greater car sharing and public commuting, further reducing the need for parking spaces and optimizing automobile usage.[270] Finally, our environmental footprint would be reduced due to both a reduction in energy demands and a transition to greener automotive energy alternatives.[271]

An important step towards the future of transportation and the auto industry is the electric car. Successful efforts by Tesla, Toyota, Nissan, GM, and others[272] in increasing battery range, establishing a charging infrastructure, and reducing prices will lead to wide-scale adoption of this technology within a decade. An important added benefit will be the reduction of our carbon footprint. While the performance and range of electric cars in the future will be continuously improved,[273] new transportation technologies are envisioned that build on the electric car platform with a focus on autonomous capabilities. Many components of autonomous functionality have already begun to be available in automobiles, quietly introducing semi-autonomy to the public. For example, Tesla's semi-autonomous mode includes features such as AutoSteer, Traffic-Aware Cruise Control, Auto Lane Change, Automatic Emergency Steering, and AutoPark,[274] allowing for hands-free driving on the highway.

The concept of driverless vehicles sparks a significant amount of interest in our convenience-focused culture in the US. The promise of providing greater safety while simultaneously requiring little to no attention to the process of driving is seen as a win-win for consumers. "Imagine eliminating 9 out of every 10 car crashes. That's the bright promise driverless technology holds over the next 30 years."[275]

Companies are now experimenting with autonomous vehicle development and testing throughout the world, with mostly positive results. Google recorded 1.45 million miles in autonomous mode on public roads in California, Texas and other states before the first accident was caused by one of their test vehicles.[276] In April 2016 a fleet of autonomous trucks from multiple manufacturers successfully crossed Europe as part of a truck platooning competition,[277] Ford successfully tested its LIDAR-equipped Fusion Hybrid autonomous vehicle in complete darkness in the desert,[278] and in July Mercedes-Benz successfully tested its CityPilot autonomous bus technology on a 12-mile bus route in Amsterdam.[279] However, recent serious accidents as a result of the use of Tesla's Autopilot autonomous control software in its vehicles have raised concerns around the reliability of such systems.[280] Another concern is ensuring the security of vehicle systems with a continuous connection to the Internet. This is a significant challenge, as demonstrated in 2015 with the remote takeover of braking, steering, radio, windshield wipers, and climate control of a Jeep. The hackers successfully interfered with the operation of the vehicle from a house ten miles away.[281]

Beyond issues of autonomous system performance, the adjudication of liability in the event of an accident is a difficult social issue. It requires critical policy decisions to be made about how our laws will ensure public safety and culpability. In the interim, Google, Volvo and Mercedes have all made pledges to accept liability for any accident their vehicles cause in full autonomous mode.[282] Will insurance companies issue policies based on the accident history of a vehicle's make and model in lieu of driving skills and history, which are no longer relevant? KPMG studied the impact of these future changes.[283] They predicted significantly fewer vehicle purchases as consumers utilize a shared vehicle model, fewer but increasingly expensive accidents, lower overall insurance premiums, and a shift in focus to commercial insurance rather than personal insurance. As autonomous vehicles improve and accident rates drop far below those of vehicles piloted by human drivers, insurance companies may mandate that humans no longer be allowed the freedom to drive unless they pay extremely high premiums.

Companies such as Uber, Zipcar and Lyft have, to a large extent, popularized the concept of ridesharing and collective ownership of automobiles. Uber is already working on redefining the segment, developing autonomous automobiles designed to supplement other forms of ridesharing. Meanwhile, a small company called nuTonomy will roll out the first fully autonomous taxi service in Singapore in late 2016.[284] With a successful autonomous golf cart trial and the involvement of leadership from the MIT robotics labs, the company is taking advantage of the country's progressive attitude towards autonomous vehicles to set up the first trial location for its service. Much as Uber's tremendous success using on-demand drivers has resulted in a significant social impact on traditional taxicab services and their drivers, autonomous taxi services may further damage an already struggling industry. The impact of even the future promise of autonomous vehicles for public transportation is already being felt. Local governments throughout the country are questioning plans to build transit options such as new rail systems that take 15-20 years to build, as they may become unnecessary once autonomous buses and taxis are in use.[285]

The aviation industry is expected to undergo significant changes in the future by further lowering costs and providing for faster travel. One recent example of this direction was the launch of Airbus A380, the world's largest passenger aircraft, with the capability to carry nearly 1,000 passengers at 15% lower operating costs per passenger. While the A380 may have defined a ceiling in scale, future efforts utilizing new materials and processes should continue to bring fast, safe, and reliable air travel down in cost.[286]

With a predicted seven-fold increase in commercial air traffic by 2050, this form of transportation will have a growing impact on the environment at current levels of pollution. Tesla's Elon Musk believes an expected 70% reduction in the cost of lithium ion battery technology by 2025 with an associated increase in the price of kerosene jet fuel will drive the need for electric-powered aircraft. While we are waiting for the availability of electrified commercial aircraft, the adoption of biofuel for jet aircraft will result in a drop of carbon emissions of 36-85%. Bio-inspired wings are another future technology option being considered for future aircraft, with new flexible materials able to change shape mid-flight to improve fuel economy.[287]

Beyond efficiency improvements, automation is already having an impact in civil aviation, and this will increase in future. Unbeknownst to many, the only period during which the pilots control commercial aircraft is during takeoff and landing. For the rest of the flight, the aircraft's autopilot is typically engaged.[288] However, as proven by the events of September 11, 2001 in New York and Washington D.C., Malaysia Airways Flight MH-370, and Germanwings Flight 9525, the most dangerous factor in a commercial aircraft can be the human pilot. Fully autonomous airplanes are currently under development to completely remove the pilot from the equation. As the NAE notes, however, the process has many challenges, including cyber-physical security, human-machine integration, decision-making, and communications. Additionally, licensing and certification, public policy, and legal and social factors may be impediments.[289]

In late 2016, Ehang, a Chinese drone manufacturer, is scheduled to release its 184 autonomous aerial vehicle, a carbon-fiber bodied drone built for human passengers. It can fly at heights of up to 11,000 feet, transporting one passenger approximately 10 miles, at a speed of approximately 60 miles an hour. The purchase price of this drone is around $300,000. While the technology is in place, policy to allow humans aboard automated personal drones is yet to be developed, and the company faces hurdles in getting approval for flight, particularly in the US. With a simple point and click map interface and preset takeoff and landing spots, personal drones may soon take to the sky, autonomously moving the upper echelon of society from one location to another.[290]

A new and completely different concept of travel is the Hyperloop, a "train" of capsules shooting through an almost air-free tube, accelerated to 760 mph by a magnetic field.[291] A trip from New York to Los Angeles is predicted to take 45 minutes. Theoretically, one could then work on the West Coast of the US and commute back and forth each day to a home on the East Coast. Millions of people travel long distances to reach their workplace every day and would welcome the technology, but the reality of travelling in a capsule inside a vacuum tube at approximately the speed of sound clearly raises safety and security concerns. It is also reasonable to ask how the challenges of construction cost, jurisdiction, policy and other factors will be addressed for such a novel form of transportation. Even if proven to be safer than other modes of travel, public perception of safety may override these technical advantages and challenge general adoption.

4.3 Wireless Communications

The impact of wireless technology in all facets of our lives has been extraordinarily significant. Access to wireless communication technologies in developed and developing societies is now seen more as a right than a privilege, and a mobile connection to the global internet is becoming a staple of our modern human existence. In the *International Journal of Electronics and Communication Technology*, Kumar et al.,[292] describe the renaissance mobile network technologies have undergone since they appeared in the early 1980s, from the first generation (1G) mobile network system which enabled basic voice communications to the fourth generation (4G) network system that allows us to use mobile communication and related apps to their fullest with speeds approaching that of wired broadband cable lines.

A recent study by Cisco Systems showed that, as a result of the greater bandwidth afforded by widespread 4G/LTE mobile technologies, global mobile data usage was at 3.7 exabytes per month in 2015, an increase of 74% from the previous year.[293] The company further reports that the overall mobile traffic has grown 400 million-fold over the last 15 years. This number, while extraordinary, is not surprising given the combination of faster communications capabilities and the proliferation of mobile phones, with the number now exceeding the world's population.[294]

Intel Corporation, long a stalwart of the computing revolution, announced in April 2016 that it would cut 12,000 jobs as the personal computer market continues to shrink in favor of smaller personal communications devices such as smartphones and tablets. Due to its focus on CPUs rather than the less powerful system-on-a-chip (SoC) found in such devices, the company faces a difficult future. While the growing number of online services will need larger server farms powered by traditional CPUs, clearly at the user level, the SoC is the future as devices become smaller and thinner.[295]

Cisco Systems further projects that with increased usage of mobile devices, global mobile data traffic will exceed 30 exabytes per month by 2020. For reference, that is roughly equal to the contents of 6 billion DVDs, and it has been said that 5 exabytes would capture every word ever said by humanity throughout all time thus far! Moreover, by 2020 the average smartphone will generate traffic every month corresponding to a fivefold increase from 2015.[296] Nokia, one of the organizations at the forefront of developing next generation network systems to enable mobile networks, has a vision for 2020 that includes supporting 1000 times more capacity, implementing low latency self-aware networks with significantly lower energy needs.[297]

With network speed up to 10 gigabytes per second and reduced latency and energy costs, fifth generation (5G) wireless networks will play a critical role in future wireless devices. Volker Held of Nokia Networks notes that beyond cellphones, 5G will also allow for a focus on communication between the various sensor-equipped devices we will put online, otherwise known as the Internet of Things (IOT).[298] With cheaper and smaller communications chips and lower power needs, 5G IOT devices will be literally everywhere.

In a TED talk at the University of Edinburgh, Harold Haas described how wireless communication has become an important utility like water and electricity, and that the increased use of this utility requires new technologies that offer higher capacity and efficiency while operating at low energy. In order to meet these capacity needs, he suggests that the replacement of 14 billion light bulbs already installed worldwide by LEDs with a special chip onboard could create an enormous capacity for wireless communication, since light in the visible wavelength range is a free commodity.[299] This emerging technology in the space of wireless communication is *light fidelity*, or Li-Fi. Different from conventional Wi-Fi that relies on radio waves, Li-Fi refers to data transmission through light signals in the visible wavelength range but at higher frequency range using LEDs. LEDs are a low cost technology where light emission levels can accurately be regulated at high frequency, and along with point-to-multipoint and multipoint-to-point communication, can enable a complete data transfer networking matrix anywhere the LEDs are installed and within sight.[300]

Although this new technology is still in its infancy, researchers have already demonstrated speed exceeding 3 GB/s from a single cell LED.[301] From a security point of view, since light cannot penetrate walls, Li-Fi offers more security than data transmission through radio waves for interior spaces. As a complementary data transmission solution to cellular and Wi-Fi, Li-Fi also enables ubiquitous wireless hotspots in dense urban areas, wireless charging of mobile devices using solar receptors, safer Internet access for hospitals and RF-sensitive spaces, and a more efficient and lightweight airplane entertainment and Internet system. Also, Li-Fi could be used for real-time traffic updates using LED traffic lights that would serve not just to control traffic, but also as a sensor and communications beacon.[302]

Passive Wi-Fi is another possible future wireless communications technology that literally takes power from the air. Devices use a process referred to as "back-scattering" to absorb energy from radio signals to generate power for a new type of low power device that uses 1/10,000th as much power as today's Wi-Fi chips. Future cheap and tiny sensor-equipped IOT devices might use radio signals on different frequencies to power themselves and their own Wi-Fi chips to communicate their data, allowing for ubiquitous tracking devices for embedding in almost anything.[303]

The social impact of wireless communications has been and will continue to be tremendous. With more services moving online even in developing countries, the demand for wireless devices such as smartphones is growing across the globe. Wireless communications allow for global businesses from anywhere. A woman in a small village in India can sell items to a worldwide community.

In order to provide Internet access to those who live in regions where the infrastructure does not yet exist, Facebook has undertaken a project to provide access via a fleet of unmanned, high-altitude solar planes. As of July 2016, Facebook has successfully tested one of its planes, named Aquila, a 900-pound drone with a wingspan of 30 feet wider than a Boeing 737 jet. The planes are in the air for only thirty minutes during the test, but the plan is for them to remain in the air for ninety days before returning to the ground, an almost seven-fold increase over the current record for solar-powered planes. When successful, the system would use lasers to bring a high speed Internet connection to base stations on the ground throughout the world.[304] This effort is a rival to Google's Project Loon, an attempt to use high altitude weather balloons at an altitude of approximately 12.5 miles to provide Internet connectivity to areas that current cell tower installations do not service. Google's initial goal is to provide an uninterrupted circle of Internet access via these balloons in the Southern Hemisphere, then to create a global network that ensures internet connectivity almost anywhere in the world.[305] The possible positive and negative social ramifications of seamless global Internet connectivity are extraordinary, particularly in the developing world.

Next generation voice technology to control wireless devices of all types is being developed by Chinese Internet giant Baidu. With functionality far beyond the current systems such as Apple's Siri, teams from Baidu are working in Beijing, China, and Silicon Valley in the US to create and refine Deep Speech 2. This is a speech recognition system utilizing a deep learning neural network that has become more accurate than a human at transcribing Mandarin, and it can work with any language where enough sample data is made available for analysis. The hope for the system is to allow for new and existing smartphone users to more efficiently use their smartphones with only a voice interface, as well as to bring online those with low literacy in China and other countries without the difficulty of physically manipulating a device's touch interface. Andrew Ng, Chief Scientist for Baidu, shares his vision for the future, saying, "I would love for us to be able to talk to all our devices and have them understand us. I hope to someday have grandchildren who are mystified at how, back in 2016, if you were to say 'Hi' to your microwave oven, it would rudely sit there and ignore you."[306]

4.4 Medicine

With a growing and aging world population, the need for medical care will significantly increase in the future. Further, there is a continuous increase in the need for medical aid in the developing part of the world due to the increasing number of infectious disease outbreaks.[307] But the availability of modern medicine is highly geographically dependent. There is a large discrepancy between the distribution of the number of doctors for every 10,000 patients in Britain (27.4) and India (6).[308] One could extrapolate this data and consider how these numbers will change by 2050 when the world population is expected to reach almost 10 billion and conclude that the disparity will be even higher.

Currently, there are organizations in India and elsewhere that have found ways to meet this growing need for greater numbers of medical care practitioners by employing more "medical technicians" than doctors to reduce their workload and overall cost.[309] However, this solution may not be sustainable in the long run in light of the dearth of medical professionals projected in the future. Technology, and especially AI, could be an answer that can adequately address this issue.

With higher speed, lower latency networks and greater use of robotic surgery systems like the Da Vinci robot, costs will drop, and reliable medical technology and telesurgery will be within reach of more of the population in certain countries. For the tech-savvy, there are an increasing number of devices that facilitate the maintenance of a healthy lifestyle by keeping track of personal health metrics. There has been a recent boom in the number of devices on the market that combine sensors to measure a variety of personal health data. Most connect wirelessly to a smartphone and enable users to automatically maintain simple health metrics to provide a glimpse at an overall picture of health. But these devices are not currently advanced enough to provide much diagnostic or predictive capability. In the future these devices will be able to incorporate a wider range of sensors at a smaller size, combined with longer battery life, lower cost, and direct communication with powerful future AI-driven systems to provide significant actionable predictive health assessment and diagnosis.

For those who want DNA analysis, companies such as 23andMe currently offer services. They provide genetic analysis reports obtained from DNA saliva studies. The saliva is provided by the customer in a collection kit from 23andMe. The information received by the consumer is FDA-approved, and suggests only percentage likelihoods of developing certain medical conditions later in life based on DNA markers. In the future it is possible that decoding the human genome will become a cheap, trivial pursuit performed almost instantly and perhaps even encouraged by doctors for the young. Clearly, the impact on society could be significant.

As noted by Vivek Wadhwa, an American technology entrepreneur, and confirmed by anyone who has played "Dr. Google" and searched for medical symptoms online, the future of medical consultation is gravitating towards the Cloud. In his article "The Future of Medicine Is in Your Smartphone," Eric Topol describes how smartphones and other digital technology will revolutionize the medical field through smart solutions, further enabling self-diagnosis without the need for a visit to the doctor's office.[310] An example of a future mobile application is the tracking and collecting of comprehensive data on the environment (air quality, pollution levels, allergens, etc.) This is particularly useful for individuals suffering from severe asthma or COPD, who can modify their medications and/or elect to avoid dangerous exposure to poor air quality.[311] ResApp, an Australian based startup, recently developed an app that can determine respiratory illnesses with up to 96% accuracy just by analyzing your cough, recorded on your personal smartphone.[312]

The future of online health applications is very broad, and there is uncertainty caused by a number of problems. Some of these are extensive data-sharing between organizations, personal health data security, health policy, and clarification and relaxation of rules for what can and cannot be done with the available data, health researchers struggle even when dealing with relatively simple data today. These issues will also impact society in adopting future medical technologies such as targeted drug delivery, DNA engineering and gene modification, personalized assistive technologies, organic prosthesis, and synthetic organs.

The ability to design and manufacture synthetic organs, customized prosthetics, and implants will be greatly enhanced by the greater use of 3D printing or additive manufacturing. C. Lee Ventola discusses the current chronic shortage of human organs for transplantation. In the United States alone only 18% of the patients waiting for an organ donor in 2009 received an organ, while nearly 25 patients died per day waiting for an organ to arrive.[313] Another major factor is that many cannot afford the high costs of organ transplantation, estimated in 2011 at an average of over a quarter of a million dollars for a kidney to over one million dollars for a heart-lung transplant.[314]

If we could design and 3D print synthetic replacement organs tailored to the tissues in a patient, these theoretically lower-cost organs and implants would allow the economically disadvantaged to also benefit from this technology. In addition to high precision and personalized design that best meet the requirements of the individual, some of these artificial organs will not age or degrade with time. Other areas where medical applications of 3D printing have made major progress and will play an important role in the future include artificial stem cells,[315] synthesis of human skin,[316] surgical tools, and personalized medicine.[317] The manufacture of prosthetics shows particular promise in developing countries, where

> *Making a prosthetic limb the traditional way takes about a week. First, the prosthetist takes a plaster mold of the end of the damaged limb. Then the prosthetic needs to be built and sized manually. That's a lot of labor to meet the needs of the world's more than 10 million people with amputations or congenital limb damage. The problem is compounded in developing countries, where most of the amputees reside, because prosthetists are in very short supply.*[318]

The future of 3D printing of personalized medicine became a reality after Aprecia became the first pharmaceutical company to receive Food and Drug Administration (FDA) approval in the US to create 3D printed tablets of an epilepsy drug in 2016. Wake Forest University researchers showed the ability to customize a 3D printed pill's content mixture and properties for an individual user allows for much more precise treatment, eschewing the one-size-fits-all approach currently used.[319]

AI will have significant applications to medicine. Wadhwa discusses the ability of systems like IBM's Watson to more accurately predict and diagnose cancer and provide advice on relevant treatments based on latest medical advances more efficiently than human doctors.[320] A specific example of this is IBM Watson for Oncology, which provides treatment options based on expert training.[321] Immune engineering or immunotherapy shows promise in future cancer treatment by engaging in engineering at the cell level – using gene-editing to modify T-cells that normally fight infection in the body and allowing them to attack cancer cells. Three hundred patients have been treated with a high rate of success since 2011 using this method to fight leukemia, and many companies are now developing treatments for cancers and even HIV using edited T-cells. Billions of dollars are being spent on these treatments, with Pfizer's head of biotechnology John Lin noting, "We think that this fundamental principle, engineering human cells, could have broad implications, and the immune system will be the most convenient vehicle for it, because they can move and migrate and play such important roles."[322]

From a social perspective, simplified access to vast stores of medical information most distinctly makes an impact in the lives of those who do not have ready access to medical care or to those who are economically disadvantaged here and abroad. A connection to the Internet and the publicly accessible medical data contained within opens up a world of medical information to a small community doctor in a way unlike ever before in human history.

4.5 Big Data and Crowdsourcing

Our world is gradually gravitating towards autonomous societies, and increasing aspects of our lives are being digitized and stored as data. As a result, massive amounts of data will continuously be produced and stored in the Cloud, and with the growth of the IOT, the use of this data will see exponential growth.

Areas where we will leave pronounced digital footprints of our lives include health and medical care, sensors, travel paths, financials, social interactions, food consumption, entertainment, business and work, shopping, e-learning, and more. One interesting question and challenge is how people/society will consider and treat each other in light of their public digital footprint, and will there be any opportunities to recover and/or erase these footprints? Certainly, businesses will greatly benefit from the analysis of vast stores of data relating to their enterprise. While businesses will be able to track the likes and dislikes of their consumers and thus package and present their products accordingly, they will also be able to use a variety of data sources internally and externally to track private data about employees, even predicting the likelihood of pregnancies and future health. While the confidentiality of personal health-related information is currently protected under the Health Insurance Portability and Accountability Act of 1996 (HIPAA), enterprising companies are using search queries and other information online to build employee profiles from correlations in the data, since there is almost no law that controls the way big data may be used.[323]

Governments will also be able to better track members of society, and thus more efficiently use and distribute their resources and "enhance safety." Although this may sound very positive overall, and result in improvements in cost effectiveness when deploying government resources, there are many privacy concerns that need to be carefully considered. From an engineering standpoint, one challenge with big data will be to design technologies that enable sufficient data generation, processing, and storage capabilities both in terms of infrastructure as well as applications. Moreover, a greater challenge will be to filter factual data from false data as well as to establish appropriate security mechanisms. Perhaps this is where advances in AI will benefit future engineers dealing with big data, as it will enable accurate machine learning and analysis of unstructured data as previously demonstrated by IBM's Watson.

The collection of data from a particular group of people for a task has received recent attention. The approach is called "crowdsourcing" and is a process whereby input is received from a large group of online participants. This data helps the technology design process, reducing the common divide between corporations and society.

The website and crowdfunding company Kickstarter illustrates that customers and ultimate end users for a product react positively to being "in" on the development stages of a product from a category in which they are interested. An example of a successful platform using crowdsourcing is the XPRIZE. In their own words, "XPRIZE is an innovation engine. A facilitator of exponential change. A catalyst for the benefit of humanity."[324] The world's most significant and influential large-scale open competition group, XPRIZE kickstarted the commercial race to space with a $10 million prize for any organization that could demonstrate two successful sub-orbital spaceflights within a two week period. They may be catalysts for future prizes of even greater scale and importance, and have certainly verified the efficacy of the prize model, currently running prize competitions in the millions of dollars for lunar travel, artificial intelligence, carbon dioxide transformation, mapping the ocean floor, software to power youth learning worldwide, portable health diagnosis, and adult literacy.[325] Other examples of crowdsourcing include the GE Open Innovation,[326] Quirky,[327] Proctor and Gamble (P&G) Connect + Develop,[328] US Department of Defense (DoD) Laboratory Innovation Crowdsourcing (LINC),[329] and OpenIDEO which is "a global community working together to design solutions for the world's biggest challenges."[330] Crowdsourcing in the future may not be purely a human-directed task, as shown by a hybrid crowdsourcing/machine process where machines organize projects crowdsourced to human labor:

> *A team of researchers at Carnegie Mellon University and Bosch argue that computers can guide the process and do a better job than humans could in many cases. In a pair of papers presented this week the team explored the concepts of using a computational system to manage the oversight of a complex project through the small contributions of individuals and using a hybrid approach that combines human judgment with machine algorithms.*[331]

As future technologies will predominantly be designed to best address the needs of the individual user and to best serve society, crowdsourced design is a platform that will play an imperative role. The impact of crowdsourcing in design will be even more pronounced in the developing world, where vast populations cannot afford or effectively use modern technologies due to economic disadvantage, lack of governance, and insufficient infrastructure. With freelance designers, engineers and local community members co-designing technologies with the ultimate user and their economic and social background in mind, future products and services will be more sustainably implemented, more readily adopted by local cultures and more sustainably implemented.

4.6 Climate Change

The world's energy consumption is predicted to grow by 56% over current levels by 2040. This is unsurprising given our growing global population and associated development needs. Although renewable energy and nuclear power are today our fastest growing energy sources, with an annual growth rate of 2.5%, fossil fuels remain our primary energy, supplying nearly 80% of the world's energy.[332] However, since their reserves are finite and they pose a threat to the environment, it is not considered viable to rely on fossil fuels in the long term. In recognition of this, at the December 2015 COP21 climate meeting in Paris, representatives from 195 member states of the UN agreed to take joint action in reducing their carbon footprint to impede the progress of global warming.[333] Climate change and energy concerns will redefine our infrastructure as well as the manufacturing industry and how we operate on a daily basis. Undoubtedly, engineers will play a key role in helping the world attempt to comply with the climate commitments made at the COP21 meeting, while simultaneously addressing our growing future needs for energy.

So, what will the society of the future look like as a result of our efforts to address climate change? We will see an increase in collective ownership of green vehicles and the rise of a society where the density of electric vehicle charging stations may surpass gas stations. Collective commuting will be preferred and facilitated by government and local municipalities, as energy and sustainability will become key parameters in infrastructure and urban planning. Optimization of traffic systems in urban areas will greatly benefit from future sensor technologies and autonomous vehicles.

With intelligent systems we will be able to significantly reduce our waste production from manufacturing, developing smarter packaging made with environmentally friendly materials and avoiding excess production. Examples of important efforts towards improving our waste management include reduced use of paper as more information is digitized and stored online and accessed only on screens, use of biodegradable polymers,[334] harvesting of energy from organic waste, and a combination of improved product design and efficient mechanics to sustainably recover reusable materials from e-waste.[335] In developing countries, however, the challenge is much greater, as when millions of people live in poverty and lack access to even the basics of human existence, sustainability and waste are not considered culturally important. This is and will continue to be a significant problem.

Water and energy are two of our most important commodities for life and are today not available to all. One in ten people do not have access to clean water,[336] and over a billion people in the world do not have access to electricity.[337] While there are many projects across the globe in making water and energy available to all, the future will see even greater engagement from other important socioeconomic stakeholders such as corporations, NGOs, and global investors. An example is Manoj Bhargava, creator of the 5-Hour Energy drink, who has established Billions in Change as "a movement to save the world by creating and implementing solutions to the most basic global problems – water, energy and health."[338] They are currently employing a group of talented engineers in the development of the Rainmaker, a seawater purification system, and the Free Electric hybrid stationary bicycle for human powered electricity generation, among other projects.[339]

In Iceland a project is under way to realize the dream of carbon capture and sequestration as part of combatting climate change. As reported in the journal *Science* in June 2016, a novel approach is being used to dissolve carbon dioxide into water, then to pump the water into rocks, locking away the CO_2 into a mineral called calcite. Volcanic basalt rocks, common in Iceland, are particularly adept at capturing CO_2 using this method, as they contain specific minerals that react with the gas in the water. After two years, 95% of the CO_2 injected into rocks had turned into calcite, safely encased with the rock.[340]

Although there is some disagreement on the efficacy of using geoengineering for climate control, several ongoing efforts in geoengineering may have a large-scale impact on localized climates. In 2016 the United Arab Emirates (UAE) offered an international research prize of $5 million for ideas in weather modification as part of their efforts since the 1990s to increase rainfall in the chronically arid country. Over the next few years, they will use nanoparticles to seed storm clouds to increase water condensation within the clouds, and therefore improve rainfall amounts.[341] The process of cloud seeding is already regularly used for climate modification around the world, in Australia, India, the US, Canada, and 48 other countries to either enhance or avoid rainfall, to mitigate hailstorms, or to clear pollution.[342] Some geoengineering concepts to at least slow, if not reverse, climate change have been suggested. One such concept involves the reflection of sunlight away from the earth by pumping sulfate aerosols in the stratosphere.[343]

Another important effort seeding future green energy technologies is the "Breakthrough Energy Coalition"[344] initiated by Bill Gates in 2015. This brings prominent global investors together to invest in future energy innovations addressing climate change, in a similar manner as venture capitalists funded successful IT start-ups in the 1990s. Although this consortium is still in its development stage, it shows promise for inspiring the next generation of young innovators and engineers who may develop new ideas of how to effectively address climate change.

In the future, the manufacturing industry will benefit greatly from many of the technological advances discussed in previous sections. The use of both semi- and fully-autonomous industrial machines will enable higher production rates, lower power consumption, and reduce waste and pollution. The development of smarter analytical tools will enable organizations to collect data at every stage of a product's lifecycle and of an organization's overall activity, including its supply chains. Technologies will be used such that the environmental footprint of the manufactured product will be minimized. An interesting example of this is the upcoming 5G wireless communications network which will reduce network energy consumption overall, and also achieve almost zero network power consumption whilst on standby mode. Along with higher data transfer rates, this will reduce end user energy needs. While insignificant on a personal level, when scaled to billions of devices this may have a discernible impact on energy needs.[345]

A timely example of future climate change-friendly energy production technologies is the development of photocatalysis, which is the light-induced acceleration of a reaction in the presence of a photocatalyst (a photosensitive semiconductor). This is similar to how plants use chlorophyll and sunlight to convert water and carbon dioxide into oxygen and glucose. One important feature of this technology is the ability to decompose water molecules to produce hydrogen gas (i.e. sunlight + $2H_2O$ [through a photocatalyst] $\rightarrow 2H_2 + O_2$, with O_2 being the only bi-product produced), which is a key ingredient in producing zero-emission fuel cells at large scale and low cost. Clearly, the ability to produce solar-derived hydrogen (in lieu of decomposing hydrocarbons that otherwise contribute to carbon emission into the atmosphere) opens up many exciting opportunities for low cost and efficient applications of hydrogen based fuel cells.[346]

4.7 Robotics

Progress in communications, AI and autonomy will play an important part in the development of next generation robotics, which will become an integral part of our lives. Robots in the future will be ubiquitous. They will be utilized in manufacturing, transportation, home help, assistive care, humanitarian care, telemedicine, unexploded mine removal from warzones, for establishing communication hubs, in search and rescue missions, in situations where accessibility is limited due to the risk of infection or radiation, and many other ways. Our overall investment in robotic systems will increase significantly in the future, from $15 billion in 2010 to an estimated $67 billion in 2025.[347] The role robotics will play will largely depend upon local culture, as well as the labor markets and the economies of the user countries. While a business owner may argue that implementation of autonomous systems and robotics will improve factory efficiency and thus profit margin, a factory worker may show deep concern, since after years of competition with cheap labor in the third world, they now also have to fight with robots over employment.

Clearly, as future technologies fall in price their impact on economically disadvantaged and developing communities that cannot otherwise benefit from technological advances will become more pronounced. This is particularly true for low cost production to support local economies, medical assistance, or humanitarian work. As illustrated by Marguerite McNeal in an article in *Wired*, robots are rapidly developing the advanced skills needed to replace their human equivalents in the workplace. One example of this is an autonomous device from Universal Robots that not only performs human tasks such as soldering and painting, but also performs "self-maintenance services" while under operational mode.[348] Another is the Knightscope K5, a 300-pound autonomous robot now being deployed as a security guard at a small number of locations throughout California. The machines are designed to replace human security guards, and with rental pricing at $7 an hour, they are an attractive option for low cost 24/7 security. They are equipped with a variety of sensors, including a thermal camera, 360-degree high definition video cameras and microphones.[349]

One distinct difference of previous advances in technology, such as those that replaced human labor in the agricultural sector, is that they were designed to address specialized skill sets and were often limited to one industry, whereas the technology of today is very different. It is broadly based and often general purpose in its application. Does this mean that the workers may not be able to transition to other industries like they did previously, because robots or the current technology will scoop up those jobs as well? [350] How will current and future generations cope with such an automated economy? Author Martin Ford says that incentivizing education with guaranteed income will be the best way to mitigate the problem. He says that this solution will make the population more entrepreneurial due to the sense of stability offered by the guaranteed income, and consequently the economy will flourish. In support of this view, he refers to the "Peltzman Effect," which states that people are usually willing to take more risk when protected by a safety net, and this could be applied to economics as well.[351] Alastair Bathgate, CEO of Blue Prism, believes that the impact of future robotics technology will not only affect blue-collar jobs but leave pronounced footprints across all professions. Similar to Ford, Bathgate believes that strong competition with robots will move people to more intellectually challenging tasks.[352]

When discussing the future of robotics, many think of the robotic vacuum cleaner Roomba from iRobot,[353] or Sony's robotic pet Aibo, which soon after it was first released in late 1990s became so popular in Japan that some "pet owners" held funeral services for them when they eventually broke down.[354] AI and deep learning have broadened the horizon of what future robots will be capable of. A good example of this is IBM's Watson, a system that not only does what it is directed to, but processes and evaluates the information, based on human-like reasoning, before coming to a conclusion. Thus, future robotic systems will not only offer us services that will facilitate our everyday lives (e.g. cooking, cleaning, replacing a broken light bulb, or buying groceries), they will change the way we humans communicate with each other as well as our overall behavioral patterns in society, since the machines will be given the equivalent of human intellect.

A Pew Foundation study of US societal views on future technology in the next 50 years indicates that there is clearly a concern about the direction technology will follow. The overall response from the participants in the survey on the impact of technological development on society was positive, with 59% expecting a better life in the future from improved transportation and medical systems. However, 65% expressed concern about the use of robots to provide support to the elderly.[355]

The future of robotics in health care is particularly interesting since it involves telemedicine and long-distance surgeries. In the future, we may have surgical centers established in any part of the world, and surgeons could use them to perform surgeries in remote locations without having to be physically present with the patient. This would enable quick action in case of emergency surgeries without the traditional challenges of travel time and accessibility of the surgery location, as well as reductions in cost.

In 2001 "Operation Lindbergh" was the codename given to the first long distance operation performed using robotic telesurgery between New York and France. This was successful, even with the significant delay in the communication of steps. However, with increases in communication speeds and lower latency connections in future, it is not unreasonable to imagine much greater use of remote surgery in this manner for specialized operations around the world. "When Mehran Anvari picks up a surgical instrument and cuts into somebody's flesh, he doesn't use his own hands. In fact, he's not even in the room. He operates on patients that are 400 kilometres away."[356]

The benefits of robot-assisted surgery are still under contention. These surgeries are currently expensive, and the benefit of using the robot instead of traditional methods is not always clear, although a study conducted by the University of California, Los Angeles (UCLA) found that a person who has undergone robot-assisted prostate surgery is less likely to require further radiation or hormone therapy compared to those who have standard "open" surgery.[357]

Moving beyond simple robotic procedures, the STAR, or Smart Tissue Automated Robot performed the first autonomous soft-tissue surgery in May 2016. The system outperformed all other forms of surgery for an intestinal anastomosis procedure – even robot-assisted human surgery.[358] However,

> *Before autonomous surgical robots can be used in places as diverse as developing countries or even in space, the researchers need to make sure the tools can be used for other soft-tissue procedures, such as removing tumors or gallbladders. The researchers believe this is fully possible – they anticipate that it could make its way to the clinic in the next two to three years.*[359]

Presently, most surgical operations are performed at a scale commensurate with the ability of trained human eyes and hands. However, it is possible to provide patients significantly more precise surgical procedures when the scale of the operation is shifted to microscopic levels. Microscopic surgery is not radically new, but the number of people who can effectively deliver it is very limited. Future advancements in robotics will help surgeons to operate at microscopic levels. The robotic arm could be made to mimic the motion of the surgeon's hand at a microscopic scale, enabling surgeons to effectively operate the targeted region. This idea is analogous to how an image editor performs various tasks on individual microscopic pixels of an image in Photoshop simply by using the mouse connected to the computer.[360]

Micro Grippers are mini robots developed by David Gracias and his team at Johns Hopkins University. Micro Grippers are star-shaped and are about 1 mm in size from tip to tip. These star-shaped micro robots are temperature sensitive, and when they are injected into the body, they respond to the body temperature by gripping the body tissue. Their gripping action can be made sensitive to other biological factors such as pH levels and the presence of certain enzymes. The retrieval of these robots from the body can be accomplished using magnets.[361] They have already been tested on animals to perform automated gastrointestinal tract biopsies, this being the only procedure they are currently suitable for due to easy access through openings in the body. Significant future improvements in design and scale are required to make this technology suitable for use in the brain, eyes, blood vessels, etc. However, researchers believe that the potential uses of these microrobots for biopsy and other purposes are vast.[362]

A different type of robot is Unmanned Aerial Vehicles (UAVs, or drones), which offer a variety of capabilities for disaster relief, surveillance and transportation. They can enable extensive reconnaissance of disaster-stricken areas, enable communication in such areas, and also deliver necessities via the air if accessibility in the area is challenged.[363] In the future, UAVs with onboard robotic actuators, a wide array of sensors, and high power lightweight batteries will revolutionize search and rescue operations. The mortality factor of humans limits their viability in dangerous and inaccessible locations, and this has led several researchers to incorporate AI and robotics in an effort to increase safety. Additional ways robotic systems may be of help in a disaster are to establish wireless communication hubs and for the remote delivery of food and medical aid to survivors.

The recent shooting of five police officers in Dallas, Texas, and the ensuing gun battle between police and the suspect was described as having turned the streets into an urban warzone. In a first for a civilian force in the US, police used a robot-delivered bomb to protect officers from further harm, detonating it and killing the suspect remotely.[364] It is expected that this type of intervention will grow in future, as military robotics makers want to get more robotic technology into the hands of the police. According to Sean Bielat, CEO of Endeavor Robotics, “That’s how robots are intended to be used.”[365] As armed military robots become cheaper, Bielat envisions a future where every police car may be equipped with a robot with armament on board, keeping the human officer out of harm’s way.

Driven by competition from new manufacturing regions with lower wages, a movement has begun toward having "dark factories." These are entirely automated factories where products are assembled robotically and humans are not required. In China, where a quarter of the world's products are manufactured, the government and private industry are putting billions of dollars into factory automation. The CEO of a Chinese company developing automated factory technologies summarizes the situation as follows: "It is very clear in China: people will either go into automation or they will go out of the manufacturing business".[366] But where will the dark factory of the future leave the millions of Chinese employed in manufacturing? Clearly, a sudden shift towards greater automation will result in economic hardship and social unrest.

Society's acceptance of robotic involvement in manufacturing may be tested on a very personal level, as the future of meat cutting and preparation may also become the work of a robot. These jobs require manual dexterity, skill, and sense of touch and typically have been considered immune to automation. However, two research groups are now testing systems that use 3D vision to analyze a chicken and then debone it as quickly as a human.[367]

Momentum Machines has developed an automated burger robot, serving up a burger from freshly prepared ingredients in ten seconds from start to finish. The company estimates $135,000 a year on average is spent to employ burger cooks at fast food restaurants, and their machine's speed, small footprint, as well as reduced liability may well impact the millions of fast food workers in the US. According to cofounder Alexandros Vardakostas, "Our device isn't meant to make employees more efficient....It's meant to completely obviate them."[368] In the future, very few jobs in processing and manufacturing will truly be "safe," unless they involve perhaps managing the robotic systems in use.

The fictional television show *Humans* portrays a future where realistic humanoid robots serve humans in an everyday capacity much as a servant would.[369] With 841 million people in the world aged 60 or older as of 2013 and projections of growth of the 60+ segment of the global population to 2 billion by 2050,[370] the future need for robotic caregivers seems clear. "Traditionally among the poorest paid of the workforce, carers are an ever more scarce resource. Policy makers have begun to cast their eyes towards robots as a possible source of compliant and cheaper help."[371]

Even though those in the US may be apprehensive of the idea of robotic caregivers or companions, in Japan, they are already being utilized in pilot efforts. The government projects that due to an aging population, 2.5 million caregivers will be needed by 2025, an increase of 800,000 from current levels. While not yet widespread, Zenkokai, a social welfare corporation in the country, has developed five different robots for use in three of their eldercare facilities.[372] Japan's problem is perhaps the most acute globally, since the elderly are living longer, with 25% of the population over 65. Traditional family structures are changing, with young people moving out and working rather than staying home to care for elderly family members. There is much work to be done in developing robots for care, and the robot's optimal functionality is many years away. Until then, perhaps the benefit of robots is to combat the greatest threat to the health of many senior citizens, loneliness.

4.8 Finance

Since 2007, M-PESA, a mobile phone-based money transfer and microfinancing service launched by Vodafone, has enabled an enormous number of financial transactions in African countries from those who may have traditionally operated outside of the normal financial systems. Just the transactions in Kenya in 2014 represented half of the country's GDP![373] The technology has also enabled philanthropists in Western countries to directly reach those in need. For example, the organization GiveDirectly uses M-PESA to distribute funds to individuals in Kenya with 91% efficiency. The growth of a simple and effective methodology for the electronic transfer of funds in a system largely uncorrupted by outside forces will continue to transform the capabilities of people living in developing countries.

Today, electronic payments are ubiquitous, with everything from buying groceries at the local supermarket to investing in stocks and properties on foreign soil made possible by a simple tap on your smartphone. According to a 2015 World Payments Report by Capgemini and and Royal Bank of Scotland, the volume of non-cash transactions has reached 357 billion.[374] In the US alone, retail e-commerce in 2014 accounted for more than $300 billion, with projections of $534 billion in 2019.[375] While new mobile electronic payment services such as Apple Pay or Samsung Pay are becoming more popular due to a combination of convenience and security, these systems still use traditional financial systems.

Cryptocurrencies, a new form of money, are cryptography-based, decentralized currencies that exist outside traditional banking systems where government has control. Cryptography is used via an open source protocol to control the creation and transfer of cryptocurrencies.[376] Bitcoin was the original cryptocurrency introduced back in 2009,[377] and as of July 2016 there were approximately 15.5 million Bitcoins in circulation, each valued at around $660 US.[378] There are also a wide variety of cryptocurrencies that have evolved from the original Bitcoin code. These are commonly referred to as Altcoins, with popular examples including Bytecoin, Deutsche eMark, Peercoin, Primecoin, Litecoin and Dogecoin.[379] These alternative cryptocurrencies have different features specific to each use case, and have much lower values than Bitcoin.

With an unregulated currency, a form of economic freedom can be achieved. Funds are readily transferred between accounts, even internationally, with very small to zero transaction fees and no requirement for the involvement of a middleman in the process.[380] While all of the details about Bitcoin transactions are public, the people involved in transactions are represented by Bitcoin addresses rather than with their personal information, and this privacy has considerable value. ISIS is using Bitcoin to fund terrorism activities, with pro-ISIS supporters transferring Bitcoin to the group anonymously and without government intervention.[381] Online dark-web marketplaces, where users go to trade drugs, weapons, stolen credit cards, and other illicit goods, rely on the anonymity of cryptocurrencies for trading.[382] There are, however, also positive uses for anonymous cryptocurrencies. In an examination of the role of Bitcoin in one non-profit organization's employment of Afghan women working from home, Code to Inspire founder Fereshteh Forough notes,

> *What social media did for communication, cryptocurrency promises to do for women's autonomy. In a society that lacks banks, blockchain technology like Bitcoin offers a secure, transparent way to add value....Most importantly, it affords those marginalized by the brick-and-mortar finance system a chance to participate in the economy on their own terms.*[383]

Cryptocurrencies such as Bitcoin are still in their infancy. Education, security, scalability, infrastructure and others are important issues that still need to be addressed for widespread adoption.[384] Cryptocurrency transactions do not carry any information about the parties involved in a transaction. Infrastructure needs to be designed that honors the privacy of users yet prevents cryptocurrency from being used in criminal activities. Moreover, an important challenge to overcome is the introduction of government regulations that could prevent the growth of this type of currency. In September 2015, Bitcoin was given the official status of commodity by the U.S. Commodity Futures and Trading Commission, requiring Bitcoin operators to comply with the law and regulations of the agency.[385]

The future social impact of cryptocurrency may lie not in the use of currencies such as Bitcoin, but rather the blockchain technology behind it. The blockchain is a distributed public ledger that typically tracks cryptocurrency transactions, but companies are now investigating how to leverage it for other applications. The NASDAQ is exploring using blockchains to track stocks, a network called Ethereum is using blockchains for digital contracts,[386] and some suggest they could be used in the future for securities settlements, tax collection, voting,[387] mortgage deeds, and even marriage contracts.[388]

4.9 Virtual and Augmented Reality

Virtual Reality (VR) is defined as "an artificial environment which is experienced through sensory stimuli (as sights and sounds) provided by a computer and in which one's actions partially determine what happens in the environment".[389] Augmented Reality (AR) is a blending of real and virtual reality where computer-generated elements are overlaid (or augmented) over a camera view of the real world. After many years of development, both of these technologies are beginning to enter the mainstream, primarily due to the technological innovations of higher quality, cheaper screens and smaller SoC-based hardware from the smartphone industry.

Released in 2013, Google Glass was the first popular and widely available AR device. It was also considered to be a failure, an over-reaching social experiment that struggled due to widespread privacy concerns from the system's small size, portability and always-enabled camera.[390] Designed primarily for indoor use, Microsoft's new HoloLens system is a unique AR device that utilizes high definition holograms projected into clear goggles. The head-mounted display technology brings 3D designs off the computer screen and into the real world, facilitating natural group interaction with a shared and digitally augmented experience. Companies are currently working on developing software for this technology. Future applications are expected to include enhanced collaborative engineering design, real-world visualization of actual-size architectural designs, and visualization of other planets.[391]

A current AR example is Niantic's Pokémon Go mobile game, which gained 20 million users in just two weeks after it was released in the US on July 6, 2016. This AR game involves users collecting virtual characters in a real world setting using their smartphone. The players follow real world street maps on their phones to visit specific real world locations, then their smartphone camera turns on and they attempt to 'capture' a virtual character at that GPS location. During this period of mass adoption of one of the first successful AR mobile games, there have been both positive and negative effects. On the positive side, the game's integration with GPS and reward system for movement is resulting in much greater numbers of young people becoming physically active, and large social groups are gathering around playing the game in popular areas. On the negative side, players have been so engrossed in their experience that they have been robbed, had car accidents, and fallen off an ocean bluff.[392] In one location in Australia, crowds of a thousand people or more descend on a neighborhood each night to play the game, bringing uncontrolled noise, traffic, and waste to the area.[393]

> *Yes, Pokémon may be a game for teens and millennials, but it has irrevocably changed societal expectations of what information is presented and how it is accessed. Very soon, visitors at New York City's Museum of Modern Art will demand to know everything about the exhibit's installations far beyond what's posted on the walls. At boutique galleries in Soho, shoppers will expect to hold up their phones and obtain more details about the process on how the latest ceramic vase is handcrafted. Young women may no longer have the patience to line outside fitting rooms at H&M, but could opt for an immediate AR fitting trial. Having to type a search term into one's browser feels so yesteryear. It will be this shift in public expectation that nudges companies from all sectors to aggressively invest in AR.*[394]

Beyond AR experiences, virtual environments allow people to have experiences they cannot have in the real world. Psychological testing of VR at the Virtual Human Interaction lab at Stanford University is examining how our brains are easily fooled and not yet able to easily distinguish between a physical and virtual world. "We still have a part of our brain that tells us yes, this is real – it looks real, it sounds real, it feels real. And therefore my body is going to react as if it's a real situation."[395] A common demonstration, known as the ledge experiment, involves putting people inside a space where they are asked to step off a plank that virtually hovers over a steep drop.[396]

> *Software worker Erin Bell inched across a wooden plank suspended over a deep, rusted pit. When a Stanford University researcher asked her to step off, she wouldn't do it. In reality Ms. Bell was walking on a carpet with a virtual-reality headset strapped to her face. "I knew I*

> *was in a virtual environment," she said later, "but I was still afraid."*[397]

A common refrain of those developing VR experiences is that "your mind is wherever we tell you it is." The improving technology behind VR with a high level of interfacing into our multiple senses allows for easy manipulation of end users. Virtual software experiences may be tailored from person to person much as our Google searches currently are, and our reactions and micro behaviors in response to various situations in VR space will be carefully measured and reported to companies. Within such engaging environments, we will be more open to manipulation.[398]

Many people in developed countries in the future will be able to 'escape' their reality with varying results. Reading books or watching movies are healthy forms of escapism, but an overdependence on ultra-realistic virtual environments may lead to users finding themselves in a situation where they struggle to reconnect with the real world. A crash or power failure that causes a virtual reality experience to fail could possibly lead to major psychological problems. Perhaps others will unlock ideas and processes never before possible to imagine in real world experiences.

> *"Substitute your five senses for virtual input enough times — and you may begin to shed aspects of your identity you once thought fixed: race, gender, age, nationality. On the bright side, people may become more empathetic and less tribal. On the negative side, people may abandon their flesh selves, leaving behind loved ones."*[399]

The development of virtual training environments is seen as another important use of VR. The greatest benefits may come for challenging environments, where technicians are able to learn and test assembly procedures while maintaining their safety. In the real world, some assembly operations if performed incorrectly would have disastrous repercussions, such as the assembly of rockets and other energetic devices.[400] Future training environments may also have AI-driven avatar mentors, providing an intelligent, guided, autonomous training system in virtual space. VR has also been used in training people with autism to better handle simulated social situations.[401]

According to Stanford professor Jeremy Bailenson, the ability of VR to change how people think and behave will have tremendous impact in the future.[402] VR is already being used by clinical psychology researchers to teach coping skills to addicts by realistically recreating situations they face on a daily basis. Virtual environments such as buffet restaurants for overeating disorders, avatars in heroin shooting galleries for drug dependencies, and liquor stores and bars for alcohol dependencies are now beginning to be used to treat these addictions.[403] While these are currently expensive scenarios to create, as VR development increases the difficulty and cost of doing so will also drop, allowing even non-technical users to create realistic virtual scenarios.

The ability in the future to scan an environment quickly and render either individual items or the entire scene into VR or AR space will be transformative. Google, Intel and Qualcomm are developing technologies to use smartphones to scan objects and spaces for use in virtual environments. Furthermore, widespread use of technology to capture a 360-degree view and reproduce it live for VR hardware-equipped observers will exponentially increase the level of impact for the people who experience it.

> *In the future, witnesses at a major event will be able to document it with their mobile phones in a way that will allow others to step inside the scene – giving people an instantaneous understanding of the event that no video or photograph could provide.*[404]

As analyst Ben Schachter of Macquarie Capital notes, "Over the short term, there are challenges. But over the long term, we think it's going to change every industry on the planet."[405] As an effort to bring VR to the masses, the Google Cardboard project[406] has been very successful in leveraging existing technology with little added cost. The company developed a simple cardboard frame with two lenses for use with a common smartphone, strapped in front of a user's face. The New York Times has given out hundreds of thousands of free holders so far for smartphones, allowing for a simplified VR experience.[407] Widely available VR experiences can also be had on the first VR rollercoaster rides at Six Flags amusement parks. Riders are physically strapped into rollercoasters that are moving around the rails while simultaneously wearing a head-mounted VR unit providing a virtual ride that is synchronized to the physical sensations they are experiencing.[408] With the development of new consumer VR systems such as the Oculus Rift, Sony's upcoming PlayStation VR, HTC Vibe, Samsung Gear VR, and others, VR is poised to make its impact on many millions of people.

It is clear that the future will bring significant technological advances in the areas of AI, transportation, wireless communications, medicine, big data, crowdsourcing, climate change, robotics, finance and virtual reality. Some will bring positive change to society, improving and extending our lives through technology, while others, such as advanced AI-powered robots, may even challenge our definition of what it is to be 'human'. Improvements in wireless communications will enable a global network of billions of new IoT devices, and engineers will leverage intelligence gained from analysis of data from these devices to change the way we live our lives. They will also engage global crowds in their efforts to develop products and services customized for the needs and wants of society. As they do so, engineers must also be cognizant of how the technologies they are designing will positively contribute to problems such as climate change, the ultimate unintended consequence of engineering. It is unclear which technologies will dominate the landscape in 50 or 100 years from now, and which will benefit the developing world at the same level they do the developed countries. One can only hope that humanity will not be lost to a virtual reality, with intelligent robotics inhabiting the real world that we have left behind.

References

[242] IBMPhoto24/Feature Photo Services, "IBM Watson is Going to Japan", Digital Image, Flickr, February 6, 2015, accessed at https://www.flickr.com/photos/ibm_media/16485823162
[243] NASA/Kathryn Hansen, "Arctic Sea Ice", Digital image, Flickr, September 2011, accessed at https://www.flickr.com/photos/gsfc/6151061573
[244] Attributed to Niels Bohr in Ellis, A.K., *Teaching and Learning Elementary Social Studies*, Allyn and Bacon, p. 431, 1970.
[245] Kurzweil, R. *The Age of Spiritual Machines: When Computers Exceed Human Intelligence*, Penguin Books, January 2000.
[246] Bergen, M., "Apple is injecting iPhones with more AI to take on Google Photos", Recode, June 13, 2016, accessed at http://www.recode.net/2016/6/13/11923730/apple-ai-google-photos
[247] Kelly III, J.E., "Computing, cognition and the future of knowing: How humans and machines are forging a new age of understanding", International Business Machines, October, 2015, accessed at http://www.research.ibm.com/software/IBMResearch/multimedia/Computing_Cognition_WhitePaper.pdf
[248] "Google DeepMind", Google, 2016, accessed at https://deepmind.com.
[249] Metz, C., "Artificial Intelligence Finally Entered Our Everyday World", *Wired*, January 1, 2016, accessed at http://www.wired.com/2016/01/2015-was-the-year-ai-finally-entered-the-everyday-world
[250] Kannan, P.V., "Artificial Intelligence: Applications in Healthcare", Asian Hospital and Healthcare Management, 2016, accessed at http://www.asianhhm.com/technology-equipment/artificial-intelligence
[251] Weber, D.O., "12 Ways Artificial Intelligence Will Transform Health Care", Hospitals and Health Networks, September 28, 2015, accessed at http://www.hhnmag.com/articles/6561-ways-artificial-intelligence-will-transform-health-care
[252] Kharpal A., "How AI could make you a top stock-picker", CNBC, July 9, 2015, accessed at http://www.cnbc.com/2015/07/09/neokamis-artificial-intelligence-app-wants-to-make-you-a-top-stock-picker.html
[253] Luckerson, V., "5 Very Smart People Who Think Artificial Intelligence Could Bring the Apocalypse", *Time*, December 2014, accessed at http://time.com/3614349/artificial-intelligence-singularity-stephen-hawking-elon-musk/
[254] Kurzweil, R., "Don't fear artificial intelligence", *Time*, December 30, 2014.
[255] Luckerson, V., "5 Very Smart People Who Think Artificial Intelligence Could Bring the Apocalypse", *Time*, December 2014, accessed at http://time.com/3614349/artificial-intelligence-singularity-stephen-hawking-elon-musk/
[256] Holley, P., "Apple co-founder on artificial intelligence: 'The future is scary and very bad for people'", *The Washington Post*, March 24, 2015, accessed at https://www.washingtonpost.com/news/the-switch/wp/2015/03/24/apple-co-founder-on-artificial-intelligence-the-future-is-scary-and-very-bad-for-people/

---

[257] Gibbs, S., "Musk, Wozniak and Hawking urge ban on warfare AI and autonomous weapons", *The Guardian*, July 27, 2015, accessed at https://www.theguardian.com/technology/2015/jul/27/musk-wozniak-hawking-ban-ai-autonomous-weapons

[258] Ibid.

[259] Philips, M., "Knight Shows How to Lose $440 Million in 30 Minutes", Bloomberg, August 2, 2012, accessed at http://www.bloomberg.com/news/articles/2012-08-02/knight-shows-how-to-lose-440-million-in-30-minutes

[260] Knight, W., "Will Machines Eliminate Us?", *MIT Technology Review*, Volume 119, No. 2, p. 30., January 2016.

[261] West, J., "Microsoft's disastrous Tay experiment shows the hidden dangers of AI", Quartz, April 2, 2016, accessed at http://qz.com/653084/microsofts-disastrous-tay-experiment-shows-the-hidden-dangers-of-ai/

[262] Simonite, T., "Google's Quantum Dream Machine", *MIT Technology Review*, Vol. 119, No. 2, p. 19.

[263] "IBM Quantum Computing", International Business Machines, 2016, accessed at http://www.research.ibm.com/quantum/

[264] Metz, C., "IBM Is Now Letting Anyone Play with Its Quantum Computer", *Wired*, May 4, 2016, accessed at http://www.wired.com/2016/05/ibm-letting-anyone-play-quantum-computer/

[265] "About OpenAI", OpenAI, 2016, accessed at https://openai.com/about/

[266] Black, W.R., "Social Change and Sustainable Transport", Indiana University, January 2000, accessed at http://michaelminn.net/tgsg/publications/SCASTrpt.pdf

[267] Hummels, D., "Transportation Costs and International Trade in the Second Era of Globalization", *Journal of Economic Perspectives*, Volume 21, Issue 3, pp. 131-154, 2007.

[268] Muoio, D., "6 major ways transportation will change by 2045", TechInsider, November 17, 2015, accessed at http://www.techinsider.io/6-predictions-about-the-future-of-transportation-2015-11

[269] Corwin, S., Vitale, J., Kelly, E., Cathles, E., "The Future of Mobility Transportation Technology", Deloitte University Press, September 24, 2015, accessed at http://dupress.com/articles/future-of-mobility-transportation-technology/

[270] Hawkins, A.J., "Meet Uber's first self-driving car", The Verge, May 19, 2016, accessed at http://www.theverge.com/2016/5/19/11711890/uber-first-image-self-driving-car-pittsburgh-ford-fusion

[271] Corwin, S., Vitale, J., Kelly, E., Cathles, E., "The Future of Mobility Transportation Technology", Deloitte University Press, September 24, 2015, accessed at http://dupress.com/articles/future-of-mobility-transportation-technology/

[272] Schaal, E., "10 Car Companies That Sell the Most Electric Vehicles", The Cheatsheet, September 16, 2015, accessed at http://www.cheatsheet.com/automobiles/10-car-companies-that-sell-the-most-electric-vehicles.html

---

[273] Schaal, E., "The Price of Electric Car Batteries is Dropping Fast", The Cheatsheet, April 2, 2015, accessed at http://www.cheatsheet.com/automobiles/electric-vehicle-battery-costs-are-dropping-faster-than-expected.html
[274] "Model S Software Version 7.0", Tesla, 2016, accessed at https://www.teslamotors.com/presskit/autopilot
[275] Ziegler, C., "6 reasons to be terrified about the future of transportation", The Verge, February 2, 2015, accessed at http://www.theverge.com/2015/2/2/7966437/6-reasons-to-be-terrified-about-the-future-of-transportation
[276] Statt, N., "Google's bus crash is changing the conversation around self-driving cars", The Verge, March 15, 2016, accessed at http://www.theverge.com/2016/3/15/11239008/sxsw-2016-google-self-driving-car-program-goals-austin
[277] Moon, M., "Platoons of autonomous trucks took a road trip across Europe", Engadget, July 4, 2016, accessed at https://www.engadget.com/2016/04/07/european-truck-platooning-challenge-video/
[278] Wittmann, A., "No Lights? No Problem! Ford Fusion Autonomous Research Vehicles Use Lidar Sensor Technology to see in the Dark", Ford Motor Company, April 11, 2016, accessed at https://media.ford.com/content/fordmedia/fna/us/en/news/2016/04/11/no-lights--no-problem--ford-fusion-autonomous-research-vehicles-.html
[279] Dalton, A., "Mercedes' autonomous bus makes a landmark trip on public roads", Engadget, July 18, 2016, accessed at https://www.engadget.com/2016/07/18/mercedes-autonomous-bus-makes-a-landmark-trip-on-public-roads/
[280] Lambert, F., "Understanding the fatal Tesla accident on Autopilot and the NHTSA probe", Electrek, June 2016, accessed at http://electrek.co/2016/07/01/understanding-fatal-tesla-accident-autopilot-nhtsa-probe/
[281] Greenberg, G., "Hackers Remotely Kill a Jeep on the Highway—With Me in It", *Wired*, July 21, 2015, accessed at http://www.wired.com/2015/07/hackers-remotely-kill-jeep-highway/
[282] Harris, M., "Why You Shouldn't Worry about Liability for Self-Driving Car Accidents", IEEE Spectrum, October 12, 2015, accessed at http://spectrum.ieee.org/cars-that-think/transportation/self-driving/why-you-shouldnt-worry-about-liability-for-selfdriving-car-accidents
[283] "Automobile insurance in the era of autonomous vehicles", KPMG, June, 2015, accessed at https://www.kpmg.com/US/en/IssuesAndInsights/ArticlesPublications/Documents/automobile-insurance-in-the-era-of-autonomous-vehicles-survey-results-june-2015.pdf
[284] Ackerman, E., "nuTonomy to Test World's First Fully Autonomous Taxi Service in Singapore This Year", IEEE Spectrum, April 4, 2016, accessed at

http://spectrum.ieee.org/cars-that-think/transportation/self-driving/nutonomy-to-launch-worlds-first-fully-autonomous-taxi-service-in-singapore-this-year

[285] Morris, D.Z., “Trains and self-driving cars, headed for a (political) collision”, *Fortune*, November 2, 2014, accessed at http://fortune.com/2014/11/02/trains-autonomous-vehicles-politics/

[286] “A380: The best solution for 21st century growth”, Airbus, 2016, accessed at http://www.airbus.com/aircraftfamilies/passengeraircraft/a380family/

[287] Dove-Jay, A., “What Commercial Aircraft Will Look Like In 2050”, The Conversation, accessed at http://theconversation.com/what-commercial-aircraft-will-look-like-in-2050-33850

[288] Langewiesche. W., “The Human Factor”, *Vanity Fair*, October 2014, accessed at http://www.vanityfair.com/news/business/2014/10/air-france-flight-447-crash

[289] Autonomy Research for Civil Aviation: Toward a New Era of Flight (2014) Chapter: 3 Barriers to Implementation”, National Research Council of the National Academies, 2014, accessed at http://www.nap.edu/read/18815/chapter/5

[290] “EHANG184 Autonomous Aerial Vehicle (AAV)”, Ehang, 2016, accessed at http://www.ehang.com/ehang184

[291] “Hyperloop Alpha”, SpaceX, accessed at http://www.spacex.com/sites/spacex/files/hyperloop_alpha-20130812.pdf

[292] Kumar, A., Liu, Y., Sengupta, J., “Evolution of Mobile Wireless Communication Networks: 1G to 4G”, *IJECT*, Volume 1, Issue 1, December 2010.

[293] “Cisco Visual Networking Index: Global Mobile Data Traffic Forecast Update, 2015–2020 White Paper”, Cisco Systems, February 1, 2016, accessed at http://www.cisco.com/c/en/us/solutions/collateral/service-provider/visual-networking-index-vni/mobile-white-paper-c11-520862.html

[294] Boren, Z.D., “There are officially more mobile devices than people in the world”, *The Independent*, October 7, 2014, accessed at http://www.independent.co.uk/life-style/gadgets-and-tech/news/there-are-officially-more-mobile-devices-than-people-in-the-world-9780518.html

[295] Anthony, S., “SoC vs. CPU – The battle for the future of computing”, Ziff Davis, April 19, 2012, accessed at http://www.extremetech.com/computing/126235-soc-vs-cpu-the-battle-for-the-future-of-computing

[296] “Cisco Visual Networking Index: Global Mobile Data Traffic Forecast Update, 2015–2020 White Paper”, Cisco Systems, February 1, 2016, accessed at http://www.cisco.com/c/en/us/solutions/collateral/service-provider/visual-networking-index-vni/mobile-white-paper-c11-520862.html

[297] “Technology Vision 2020”, Nokia Communications, 2015, accessed at http://networks.nokia.com/innovation/technology-vision

[298] Held, V., “Why will we need 5G?”, Nokia Communications, December 4, 2014, accessed at https://blog.networks.nokia.com/mobile-networks/2014/12/04/this-blog-is-by-share-your-thoughts-copypaste/

[299] Waugh, R., "The future of wireless technology is coming at light speed", *The Telegraph*, January 15, 2016, accessed at http://www.telegraph.co.uk/sponsored/technology/technology-trends/12094857/li-fi-technology.html

[300] Hass, H., Yin, L., Wang, Y., Chen C., "What is LiFi?", *Journal of Lightwave Technology*, Volume 34, Issue 6, March 15, 2016.

[301] Tsonev, D., Chun, H.,, Rajbhandari, S., McKendry, J, Videv, S., Gu, E., Haji, M., Watson, S., Kelly, A., Faulkner, G., Dawson, M., Haas, H., and O'Brien, D., "A 3-Gb/s single-LED OFDM-based wireless VLC link using a gallium nitride μLED," *IEEE Photon. Technol. Lett.*, Volume 26, No. 7, pp. 637–640, April 2014.

[302] "Smart location-based guide navigation and advertising", Axrtek, 2016, accessed http://www.axrtek.com/applications.html

[303] Harris, M. "Power from the Air", *MIT Technology Review*, Vol. 119, No. 2, p. 67, February 2016.

[304] Newton, C., "Facebook Takes Flight", The Verge, 2016, accessed http://www.theverge.com/a/mark-zuckerberg-future-of-facebook/aquila-drone-internet

[305] Simonite, T., "Project Loon", *MIT Technology Review*, accessed at https://www.technologyreview.com/s/534986/project-loon/

[306] Knight, W., "Conversational Interfaces", *MIT Technology Review*, Vol. 119, No. 2, p. 43, February 2016.

[307] Smith, K.F., Goldberg, M., Rosenthal, S., et al., "Global rise in human infectious disease outbreaks", The Royal Society, October 29, 2014, accessed at http://rsif.royalsocietypublishing.org/content/11/101/20140950.full

[308] "Squeezing out the doctor", *The Economist*, May 31, 2012, accessed at http://www.economist.com/node/21556227

[309] Ibid.

[310] Topol, E.J., "The Future of Medicine Is in Your Smartphone", *The Wall Street Journal*, January 9, 2015, accessed at http://www.wsj.com/articles/the-future-of-medicine-is-in-your-smartphone-1420828632

[311] Ibid.

[312] Sadler, D., "Cough into this phone, please': Perth startup's diagnosis app could change the way respiratory problems are detected", StartupSmart, July 16, 2015, accessed at http://www.startupsmart.com.au/advice/growth/innovation/cough-into-this-phone-please-perth-startups-diagnosis-app-could-change-the-way-respiratory-problems-are-detected/

[313] Ventola, C.L., "Medical Applications for 3D Printing: Current and Projected Uses", *Pharmacy and Therapeutics*, Vol. 39, Number 10, pp. 704–711, October 2014.

[314] "Organ Transplantation: The Process", US Department of Health and Human Services, accessed at http://www.organdonor.gov/about/transplantationprocess.html

[315] Choi, C. Q., "Doctors Could 3D-Print Micro-Organs with New Technique", Live Science, December 2015, accessed at http://www.livescience.com/52942-3d-printing-embryonic-stem-cells.html

[316] Extance, A., "Cosmetics deals push skin 3D bioprinting", Royal Society of Chemistry, September 24, 2015, accessed at http://www.rsc.org/chemistryworld/2015/09/cosmetics-deals-fuel-skin-3d-bioprinting

[317] Kira, B., "How 3D printed pills are personalizing medicine", 3ders.org, January 2, 2016, accessed at http://www.3ders.org/articles/20160102-how-3d-printed-pills-are-personalizing-medicine.html

[318] "Nia's 3D-Printed Prosthetic Limbs Cut Costs and Production Time", Engineering For Change, April 18, 2016, accessed at http://www.engineeringforchange.org/nias-3d-printed-prosthetic-limbs-cut-costs-and-production-time/

[319] Kira, B., "How 3D printed pills are personalizing medicine", 3ders.org, January 2, 2016, accessed at http://www.3ders.org/articles/20160102-how-3d-printed-pills-are-personalizing-medicine.html

[320] Wadhwa, V., "Why I'm excited about the promising future of medicine", *The Washington Post*, April 7, 2014, accessed at https://www.washingtonpost.com/news/innovations/wp/2014/04/07/why-im-excited-about-the-promising-future-of-medicine/

[321] "IBM Watson for Oncology", International Business Machines, 2016, accessed at http://www.ibm.com/watson/watson-oncology.html

[322] Regalado, A., "Immune Engineering", *MIT Technology Review*, Vol. 119, No. 2, p. 34, February 2016.

[323] Edwards A., "Companies Are Using Big Data to Track Employee Health and Pregnancies", *Popular Science*, February 19, 2016, accessed at http://www.popsci.com/companies-use-big-data-to-track-employee-health-and-pregnancies

[324] X Prize, 2016, accessed at http://www.xprize.org/

[325] Ibid.

[326] "GE Open Innovation", General Electric, 2016, accessed at http://www.ge.com/about-us/openinnovation

[327] Sherman, A., "A Quirky Bid for GE's Home Appliances in More Ways Than One", Bloomberg, August 14, 2014, accessed at http://www.bloomberg.com/news/2014-08-14/a-quirky-bid-for-ge-s-home-appliances-in-more-ways-than-one.html

[328] "Connect + develop", Proctor and Gamble, 2016, accessed at http://www.pgconnectdevelop.com/

[329] "5 Simple Steps Make Linc Work", Combatting Terrorism Technical Support Office, 2016, accessed at https://linc.cttso.gov/LINC/LINC.nsf/HowItWorks.xsp

[330] OPEN IDEO, 2016, accessed at http://openideo.com/

[331] Schaffhauser, D., "Research Project Mixes Humans and Machines for Better Crowdsourcing", Campus Technology, May 10, 2016, accessed at

https://campustechnology.com/articles/2016/05/10/research-project-mixes-humans-and-machines-for-better-crowdsourcing.aspx

[332] "International Energy Outlook 2013", US Energy Information Administration, US Department of Energy, Washington DC, 2013.

[333] "COP 21 Sustainable Innovation Forum 2015", United Nations Environment Programme, 2016, accessed at http://www.cop21.gouv.fr/en/

[334] Arrighi, V., "Five synthetic materials with the power to change the world", The Conversation, February 4, 2015, accessed at http://theconversation.com/five-synthetic-materials-with-the-power-to-change-the-world-37131

[335] Szaky, T., "10 recycling and waste management trends to look out for in the near future", Treehugger, June 5, 2014, accessed at http://www.treehugger.com/corporate-responsibility/10-recycling-and-waste-management-trends-watch-out-near-future.html

[336] "Facts about Water & Sanitation", Water.org, 2016, accessed at http://water.org/water-crisis/water-sanitation-facts/

[337] "Energy Poverty", International Energy Agency, 2016, accessed at http://www.iea.org/topics/energypoverty/

[338] Billions in Change, 2016, accessed at http://billionsinchange.com/

[339] Huizinga, D., "Billions in Change Fights Poverty with Inventions", Opportunity Lives, April 8, 2016, accessed at http://opportunitylives.com/billions-in-change-fights-poverty-with-inventions/

[340] Matter, J.M., et al., "Rapid carbon mineralization for permanent disposal of anthropogenic carbon dioxide emissions", *Science*, Vol. 352, Issue 6291, pp. 1312-1314, 10 Jun 2016.

[341] Al Wasmi, N., "Academic trio with 'heads in the clouds' win $5m UAE rain enhancement research grant", The National, January 19, 2016, accessed at http://www.thenational.ae/uae/environment/academic-trio-with-heads-in-the-clouds-win-5m-uae-rain-enhancement-research-grant

[342] Little, A., "Weather on Demand: Making It Rain Is Now a Global Business – Welcome to the strange world of cloud seeding", *Bloomberg*, October 28, 2015, accessed at http://www.bloomberg.com/features/2015-cloud-seeding-india/

[343] Venakataraman, B., "What Experiments to Block out the Sun Can't Tell Us", *Slate*, January 12, 2016, accessed at http://www.slate.com/articles/technology/future_tense/2016/01/experimenting_with_geoengineering_could_have_unintended_consequences.html

[344] "Introducing the Breakthrough Energy Coalition", Breakthrough Energy Coalition, 2016, accessed at http://www.breakthroughenergycoalition.com/en/index.html

[345] "The 5G of the future: a network that will have the environment and low energy embedded in its technological DNA", Orange, November 4, 2015, accessed at http://www.orange.com/en/Responsibility/Environment/COP21/5G

[346] Kapilashrami, M., Liu, Y.-S., Glans, P.-A., and Guo, J.-H., “Soft X-Ray Spectroscopy and Electronic Structure of 3d Transition Metal Compounds in Artificial Photosynthesis Materials”, *From Molecules to Materials – Pathways to Artificial Photosynthesis*, Springer International Publishing, 2015.
[347] Thompson, M., “We’ll spend $67 Billion on Robots by 2025”, The Fiscal Times, March 6, 2015, accessed at http://www.thefiscaltimes.com/2015/03/06/We-ll-Spend-67-Billion-Robots-2025
[348] McNeal, M., “Rise of the Machines: The Future has Lots of Robots, Few Jobs for Humans”, *Wired*, accessed at http://www.wired.com/brandlab/2015/04/rise-machines-future-lots-robots-jobs-humans/
[349] Hill, K., “Uber hired a robot to patrol its parking lot and it’s way cheaper than a security guard”, Fusion, July 7, 2016, accessed at http://fusion.net/story/321329/knightscope-security-robot-uber-parking-lot/
[350] McNeal, M., “Rise of the Machines: The Future has Lots of Robots, Few Jobs for Humans”, *Wired*, accessed at http://www.wired.com/brandlab/2015/04/rise-machines-future-lots-robots-jobs-humans/
[351] Ibid.
[352] Bathgate, A., “Will Robots Save The Future Of Work?”, Tech Crunch, January 15, 2016, accessed at http://techcrunch.com/2016/01/15/will-robots-save-the-future-of-work/
[353] iRobot, 2016, accessed at http://www.irobot.com/
[354] Mochizuki, T., Pfanner, E., “In Japan, Dog Owners Feel Abandoned as Sony Stops Supporting ‘Aibo’”, *The Wall Street Journal*, accessed at http://www.wsj.com/articles/in-japan-dog-owners-feel-abandoned-as-sony-stops-supporting-aibo-1423609536
[355] Smith, A., “U.S. Views of Technology and the Future”, Pew Research Center, April 17, 2014, accessed at http://www.pewinternet.org/files/2014/04/US-Views-of-Technology-and-the-Future.pdf
[356] Eveleth, R., “The surgeon who operate from 400km away”, British Broadcasting Corporation, May 16, 2014, accessed at http://www.bbc.com/future/story/20140516-i-operate-on-people-400km-away
[357] Ackerman, E., “New Da Vinci Xi Surgical Robot Is Optimized for Complex Procedures”, Institute of Electrical and Electronics Engineers, April 7, 2014, accessed at http://spectrum.ieee.org/automaton/robotics/medical-robots/new-da-vinci-xi-surgical-robot
[358] Herkewitz, W., “Robot Performs Soft-Tissue Surgery by Itself”, *Popular Mechanics*, May 4, 2016, accessed at http://www.popularmechanics.com/science/health/a20718/first-autonomous-soft-tissue-surgery/

[359] Ossola, A., "Autonomous robot surgeon just outperformed human surgeon", America Online, May 4, 2016, accessed at http://www.aol.com/article/2016/05/04/autonomous-robot-surgeon-just-outperformed-human-surgeon/21370539/
[360] Sofge, E., "The Microscopic Future of Surgical Robotics", *Popular Science*, April 4, 2014, accessed at http://www.popsci.com/blog-network/zero-moment/microscopic-future-surgical-robotics
[361] Courtland, R., "Medical Microbots Take a Fantastic Voyage into Reality", Institute of Electrical and Electronics Engineers, June 1, 2015, accessed at http://spectrum.ieee.org/robotics/medical-robots/medical-microbots-take-a-fantastic-voyage-into-reality
[362] Gorman, C., "Miniature Robots Perform Surgery", Institute of Electrical and Electronics Engineers, June 3, 2015, accessed at http://spectrum.ieee.org/video/robotics/medical-robots/video-miniature-robots-perform-surgery
[363] Barcock, A., "Solutions That Are Saving Lives in Humanitarian Response", Aid and International Development Forum, May 8, 2015, accessed at http://www.aidforum.org/disaster-relief/top-solutions-that-are-saving-lives-in-humanitarian-response#robotics
[364] Rosenwald, M.S., "An urban war zone: 'We don't know where the hell he's at'", *The Washington Post*, July 8, 2016.
[365] Tucker, P., "Military Robotics Makers See a Future for Armed Police Robots", Defense One, July 11, 2016, accessed at http://www.defenseone.com/technology/2016/07/military-robotics-makers-see-future-armed-police-robots/129769/
[366] Knight, W., "The People's Robots", *MIT Technology Review*, May/June 2016, p. 44.
[367] Nichols, G., "Would you buy meat from a robot butcher?", The Daily Dot, February 28, 2016, accessed at http://kernelmag.dailydot.com/issue-sections/headline-story/15943/robot-butchers-future-food/
[368] Dorrier, J., "Burger Robot Poised to Disrupt Fast Food Industry", Singularity University, August 10, 2014, accessed at http://singularityhub.com/2014/08/10/burger-robot-poised-to-disrupt-fast-food-industry/
[369] *Humans*, Video, Channel 4 UK, accessed at http://www.channel4.com/programmes/humans
[370] "World Population Ageing 2013", United Nations Department of Economic and Social Affairs, 2013, accessed at http://www.un.org/en/development/desa/population/publications/pdf/ageing/WorldPopulationAgeing2013.pdf
[371] Watts, G., "The One-Armed Robot That Will Look after Me until I Die", Vice Media, April 19, 2016, accessed at http://motherboard.vice.com/read/the-one-armed-robot-that-will-look-after-me-until-i-die
[372] Katoka, K., "Robot caregivers to help elderly", The Japan News, February 23, 2016, accessed at http://the-japan-news.com/news/article/0002752218

[373] "M-Pesa", Wikipedia, accessed at https://en.wikipedia.org/wiki/M-Pesa

[374] "World Payments Report 2015 reveals accelerated growth of non-cash payments", Royal Bank of Scotland, October 6, 2015, accessed at http://www.rbs.com/news/2015/october/world-payments-report-2015-reveals-accelerated-growth-of-non-cas.html

[375] "Statistics and facts about e-commerce in the United States", Statista, 2016, accessed at http://www.statista.com/topics/2443/us-ecommerce/

[376] Carydon, C., "What is Cryptocurrency?", Cryptocoins News, September 16, 2014, accessed at https://www.cryptocoinsnews.com/cryptocurrency/

[377] Bitcoin, 2016, accessed at https://bitcoin.org

[378] "Bitcoin - Total XBT in Circulation", CoinDesk, 2016, accessed at http://www.coindesk.com/data/bitcoin-total-circulation/

[379] Altcoins, 2016, accessed at http://altcoins.com/

[380] Shin, L., "Susan Athey on How Digital Currency Could Transform Our Lives", *Forbes*, November 24, 2014, accessed at http://www.forbes.com/sites/laurashin/2014/11/24/susan-athey-on-how-digital-currency-could-transform-our-lives/#6a8738e679e7

[381] Sander IV, L., "Bitcoin: Islamic State's online currency venture", Deutsche Welle, September 20, 2015, accessed at http://www.dw.com/en/bitcoin-islamic-states-online-currency-venture/a-18724856

[382] Greenberg, A., "The Silk Road's Dark-Web Dream Is Dead", *Wired*, January 14, 2016, accessed at https://www.wired.com/2016/01/the-silk-roads-dark-web-dream-is-dead/

[383] Campbell, R., "Code to Inspire: Bitcoin Gives Afghan Women Financial Freedom", Bitcoin Magazine, April 14, 2016, accessed at https://bitcoinmagazine.com/articles/code-to-inspire-bitcoin-gives-afghan-women-financial-freedom-1460652348

[384] "What are the risks of investing in Bitcoin?", Investopedia, October 3, 2014, accessed at http://www.investopedia.com/ask/answers/100314/what-are-risks-investing-bitcoin.asp

[385] "CFTC Orders Bitcoin Options Trading Platform Operator and its CEO to Cease Illegally Offering Bitcoin Options and to Cease Operating a Facility for Trading or Processing of Swaps without Registering", US Commodities Futures Trading Commission, September 17, 2015, accessed at http://www.cftc.gov/PressRoom/PressReleases/pr7231-15

[386] Hern, A., "Blockchain: the answer to life, the universe and everything?", *The Guardian*, July 7, 2016, accessed at https://www.theguardian.com/world/2016/jul/07/blockchain-answer-life-universe-everything-bitcoin-technology

[387] "Blockchain Enigma. Paradox. Opportunity", Deloitte, 2016, accessed at http://www2.deloitte.com/uk/en/pages/innovation/articles/blockchain.html

[388] Skinner, C., "Why the blockchain will radically alter our future", BankNXT, May 8, 2015, accessed at http://banknxt.com/50938/blockchain-future/

[389] "Virtual Reality", *Merriam Webster*, 2016, accessed at http://www.merriam-webster.com/dictionary/virtual%20reality

[390] Yu, H., "What Pokémon Go's Success Means for the Future of Augmented Reality", *Fortune*, July 23, 2016, accessed at http://fortune.com/2016/07/23/pokemon-go-augmented-reality/
[391] "HoloLens", Microsoft Corporation, 2016, accessed at https://www.microsoft.com/microsoft-hololens/en-us
[392] Serhan, Y., "The Health Risks of Pokémon Go", *The Atlantic*, July 25, 2016, accessed at http://www.theatlantic.com/news/archive/2016/07/pokemon-go-health-warning/492899/
[393] Esposito, B., "Residents Are Pissed That Their Neighborhood Has Become a Pokémon Go Hot Spot", Buzzfeed News, July 12, 2016, accessed at https://www.buzzfeed.com/bradesposito/pokemon-go-rhodes
[394] Yu, H., "What Pokémon Go's Success Means for the Future of Augmented Reality", *Fortune*, July 23, 2016, accessed at http://fortune.com/2016/07/23/pokemon-go-augmented-reality/
[395] "Virtual Reality's Psychological and Behavioral Effects", Tested, February 18, 2016, accessed at https://www.youtube.com/watch?v=Cil7OT8bGik
[396] Sackman, D., "Real change through virtual reality", Video, TEDxEastEnd, February 20, 2015, accessed at http://tedxtalks.ted.com/video/Real-change-through-virtual-rea
[397] Nicas, J., Seetharaman, D., "What Does Virtual Reality Do to Your Body and Mind?", *The Wall Street Journal*, January 3, 2016.
[398] Serrano, A., "Life inside the bubble of a virtual reality world", Video, TEDxToronto, October 16, 2014, accessed at http://tedxtalks.ted.com/video/Life-inside-the-bubble-of-a-vir
[399] Carlson, N., "I Tried Facebook's Virtual-Reality Headset — It's Going To Change everything", Business Insider, December 12, 2014, accessed at http://www.businessinsider.com/oculus-is-going-to-change-everything-2014-12
[400] Gupta, S.K., Anand, D.K., Brough, J.E., Schwartz, M., and Kavetsky, R.A., *Training in Virtual Environments: A Safe, Cost-Effective, and Engaging Approach to Training*, CALCE EPSC Press, University of Maryland, College Park, 2008.
[401] Kandalaft, M.R., Didebhani, N., Krawezyk, D.C., Allen, T.T., Chapman, S.B., "Virtual Reality Social Cognition Training for Young Adults with High-Functioning Autism", *J Autism Dev Disord.*, January 2013, pp. 34-44.
[402] Nicas, J., Seetharaman, D., "What Does Virtual Reality Do to Your Body and Mind?", *The Wall Street Journal*, January 3, 2016.
[403] Bordnick, P., "How can virtual reality help us deal with reality?", Video, TEDxHouston, November 19, 2015, accessed at https://www.youtube.com/watch?v=OPfQQw72kus
[404] De la Pena, N., "Step Inside the Future", *MIT Technology Review*, Vol. 119, No. 4, July/August 2016, accessed at https://www.technologyreview.com/s/601685/step-inside-the-future/
[405] Nicas, J., Seetharaman, D.,"What Does Virtual Reality Do to Your Body and Mind?", *The Wall Street Journal*, January 3, 2016.

---

[406] "Google Cardboard", Google, 2016, accessed at https://vr.google.com/cardboard/

[407] Robertson, A., "The New York Times is sending out a second round of Google Cardboards", The Verge, April 28, 2016, accessed at http://www.theverge.com/2016/4/28/11504932/new-york-times-vr-google-cardboard-seeking-plutos-frigid-heart

[408] Franco, M., "Here's what it's like to ride a virtual reality roller coaster", New Atlas, May 25, 2016, accessed at http://www.gizmag.com/virtual-reality-coaster/43535/

Hang in there, we're almost at the end.
Please read Chapter 5.

This page is again intentionally left blank.

Bennett Hazelwood looks into the future [409]

Future engineers discussing social impact in the Engineering for Social Change course at the University of Maryland [410]

Chapter 5

The Engineer of the Future

The future is now.[411]

The engineers of the future will use their engineering skills to solve social as well as technical problems in unique and innovative ways. For those of us in engineering education, this requires a non-traditional approach. In a global market, no longer can engineering students simply be prepared to just be more technically savvy than all others; they must also have an appreciation of variations of people, place, and culture in order to understand problems far beyond the scope of the classroom. As the engineers of the future, they should be prepared to solve the world's most difficult, but not necessarily *technical,* problems that impact a large swath of humanity.

5.1 The Challenge

The global engineer of the future must have a keen awareness of his/her ethical responsibility as well as an understanding of his/her social responsibility. There is more to Engineering than just Engineering and the global engineer should be inspired to use his or her skills and mindset to practice social entrepreneurship and pursue ideas that make a difference. Our challenge as engineering educators then can be summarized by the following five "mantras":

- Inculcate an appreciation of the social change that engineering creates, and how not only for-profit enterprises, but also philanthropy and non-profits act as catalysts.
- Create engineers well-rounded in their education such that they have a global perspective, a social conscience, a better understanding of how they impact society, and a sense of their ethical and social responsibility.
- Help engineering students to understand the scope and impact of the challenges facing billions of people in the developing and underdeveloped world every day.

- Give engineers an appreciation of the important roles of public policy and philanthropy, and encourage their involvement in these areas to create positive social change.
- Encourage engineers to create innovative solutions to the unintended consequences of past engineering successes and to consider carefully the potential global impact of new technologies and products.

Recognizing that our society faces tremendous social challenges in the future, both technical and non-technical in nature, the National Academy of Engineering has defined the major challenges that engineers of the future will address:[412]

- Advance personalized learning
- Make solar energy economical
- Enhance virtual reality
- Reverse-engineer the brain
- Engineer better medicines
- Advance health informatics
- Restore and improve urban infrastructure
- Secure cyberspace
- Provide access to clean water
- Provide energy from fusion
- Prevent nuclear terror
- Manage the nitrogen cycle
- Develop carbon sequestration methods
- Engineer the tools for scientific discovery

Whether these are really the right topics is unclear. In fact, the seminal question posed by Paul Polak is: "I keep asking why 90% of the world's designers work exclusively on products for the richest 10% of the world's customers."[413] Perhaps Intel has learned this lesson the hard way, as it recently laid off 12,000 people due in part to missing the smart phone revolution that is transforming the developed and developing world.[414]

The National Academy of Engineering, currently directed by former President of the University of Maryland at College Park and Mechanical Engineering Professor C. Daniel Mote Jr., has defined the global engineer:

> *At a basic level, a "global engineer" can be defined as one who possesses the cultural and personal skills to*

> *work effectively anywhere in the world, displays outstanding technical competence, and contributes to advancing the objectives of his or her individual organization and its partners.*[415]

Clearly, a breakthrough in many of NAE's challenge areas would result in tremendous positive social change, and this is of particular importance in the developing world. Low-cost solar technology could light up many areas of the earth that are currently dark and provide power to fuel economic growth. Access to clean water could reduce illness and death in many parts of the world and may also act as an economic driver. The removal of carbon monoxide from the atmosphere could reduce the hundreds of thousands of deaths and illnesses each year from pollution and decrease growing global temperatures. Engineering improvements could not only improve the quality of life for millions of people, but also save many thousands of lives.

Naturally, it is difficult to find companies willing to work on issues with great social impact that do not also have a healthy profit margin. Humanitarian projects are generally from well-meaning people but can fail over the longer term for multiple reasons, one of which is an emphasis on the benefits to the grantor rather than those delivered to the grantee.[416] Another is failing to engage in the process of co-design, where the end user is considered an important part of the product design process. However, even with solutions where co-design is incorporated, there are often challenges to adoption. Sometimes a simple and elegant engineering solution fixes a serious social problem but fails to remain permanently successful. A case in point involves a University of Maryland team who was working on a solution to removing cholera bacteria from drinking water in Bangladesh. They discovered that if old, laundered sari cloth was first folded four times, and then used as a filter to pour cholera-laden water through, 99% of the cholera in the water was filtered out. What seems like a simple and elegant solution was very effective for a short period, but when researchers returned to Bangladesh at a later date they found many women had stopped using more than one layer of sari cloth, and some had even stopped filtering the water altogether. A holistic approach is clearly required to implement solutions that require behavioral changes, even with the threat of a fatal disease like cholera.[417] Interestingly, although the University of Maryland solution was effective, the team has struggled to get funding for community education for this very simple, low-cost, life-saving idea, while expensive and complex solutions have been taking precedence.

Engineers must be cognizant of their ethical responsibility. Institutional and political pressures will arise in an engineer's practice, and these will require a proper approach to ethics. While traditionally described as a set of rules defining the conduct of engineers as they engage in professional practice, "ethical practice in engineering is critical for ensuring public trust in the field and in its practitioners. This is especially critical as engineers increasingly tackle international and socially complex problems that combine technical and ethical challenges."[418]

The American Society of Mechanical Engineers (ASME) maintains a current code of ethics for practicing engineers:

> *Engineers uphold and advance the integrity, honor and dignity of the engineering profession by:*
>
> *I. using their knowledge and skill for the enhancement of human welfare;*
>
> *II. being honest and impartial, and serving with fidelity their clients (including their employers) and the public; and*
>
> *III. striving to increase the competence and prestige of the engineering profession.*[419]

The first point in the code of ethics notes it is the responsibility of the engineer to *enhance* human welfare, and by extension, the engineer may not *degrade* human welfare. The work of engineers in the past has unintentionally impacted society negatively in a variety of ways; so engineering students must be taught that this responsibility is high on their list when engaging in ethical practice.

In other engineering disciplines, the same responsibility is made clear by each society in their own code of ethics. The National Society for Practicing Engineers (NSPE) states that professional engineers "shall hold paramount the safety, health, and welfare of the public."[420] The American Society of Civil Engineers (ASCE) notes the same responsibility – "for the enhancement of human welfare and the environment."[421] The Institute for Electrical and Electronics Engineers (IEEE) says engineers need to "accept responsibility in making decisions consistent with the safety, health, and welfare of the public, and....to improve the understanding of technology; its appropriate application, and potential consequences."[422]

The role of the engineer is clear, but the true societal impact and consequences of the results of engineering, both intended and unintended, are not always clear. Many of them will involve serious ethical questions. Engineering decisions are sometimes made to positively serve the bottom line for a company or shareholders, but at the same time harm members of society. Consider the following two timely examples:

- A General Motors engineer covered up an ignition switch problem that was shown to result in at least 32 deaths. In his role as the lead switch engineer for GM's Cobalt and Ion models, he approved switches for production that were made below minimum specifications. After complaints, the engineer approved a new higher-torque switch but intentionally covered up both the change and the original issue. As a result, the switches in older vehicles were not replaced and began failing, and owners died in crashes that were later found to be related to the faulty switch.[423]

- Volkswagen recently admitted to installing emissions-cheating software in a number of its vehicles to meet US emissions requirements. The company blamed a group of engineers for the deception. A former company engineer, Emanuela Montefrancesco, said that an aggressive culture in the company drove engineers to compete for approval and promotion, saying "Here at Volkswagen in the last few years, we have forgotten to say, 'I won't do this. I cannot. I am sorry.'" The company has agreed to a settlement that will cost them $14.7 billion.[424]

Another recent example is Mitsubishi Motors, which inflated the tires of its cars during testing to enable the reporting of higher gas mileage to the government.[425] But this was not the first time that Mitsubishi Motors had cheated its customers. In the early 2000s, investigations found that the company had lied about defects in its cars and had covered up faults since 1977 and repaired cars secretly, instead of reporting the problems to the transport ministry. The cover-up led to huge recalls and criminal charges, and will ultimately cost the company billions of dollars.[426]

Clearly engineers have to learn to say, "I want to do this", or "I can't and won't do this", in order to appropriately fulfill their ethical responsibility. Larry Jacobson, Executive Director of the National Society of Professional Engineers (NSPE) spoke about the ethical pressures facing engineers after the BP oil spill in the Gulf of Mexico:

> *Almost all industrial processes and construction begin with the engineer who does the design. The engineer is under enormous pressure to help create profit for*

> *management, and those severe pressures influence choices — choices between the safest and most prudent design and the design that sacrifices safety in the name of cost. Lower cost usually means higher short-term profit for the company.*[427]

Do engineers have a social responsibility beyond their ethical responsibilities, as set forth by various professional societies?

> *Engineers, both individually and collectively, have not only a duty to minimize harm, but, according to the very rationale of their profession, a duty to do good. If doing good is intrinsic to engineering, then the realization of that good inescapably involves a commitment to understanding and having an important voice in shaping the social agenda; and this social responsibility belongs irreducibly both to practitioners and to the profession. Awareness that social responsibility resides in both these locations should be part of the educational process for engineering students, and should play an important role in the everyday activities of practicing engineers as well as in the activities of professional engineering societies.*[428]

Furthermore, the moral and social responsibility in the practice of engineering is a shared burden resting not just upon the shoulders of the engineer, but also a number of other entities, including Government, Non-Government Organizations (NGOs), Nonprofits, Industry, Professional Societies, and Universities. Ursula Franklin, a metallurgist and Professor Emeritus at the University of Toronto, espouses a mindset of "justice, fairness and equality in the global sense." Her recommendation for engineers is to ask seven questions of an engineering project. Does the project:[429]

- Promote justice?
- Restore reciprocity?
- Confer divisible or indivisible benefits?
- Favor people over machines?
- Minimize or maximize disaster?
- Promote conservation over waste?
- Favor reversible over irreversible?

A new focus for engineering is emerging at some universities around the country, where the idea of humanitarian and development engineering has come into vogue. This brings engineering knowledge to bear on issues impacting the lives of people living in developing countries (see Appendix for a survey of these educational institutions). Bill Gates Sr. of the Bill and Melinda Gates Foundation, in speaking to engineering students, characterizes the engineer's growing social responsibility:

> *From my office at the Gates Foundation, I don't have the foggiest idea how to help Bolivian villagers. But when you visit Bolivia, spend time in people's homes, and feel the sting in your eyes from the stove fumes, you figure out pretty quickly that those fumes – and those stoves – are dangerous....being a good engineer, or a good doctor, or a good parent, or a public-spirited citizen, now encompasses engagement with the whole wide world that surrounds us. That world must become, and it is becoming, a part of our consciousness. And I don't believe for a second that we can be conscious of suffering and still do nothing about it.*[430]

5.2 Engineering for Social Change Course

At the University of Maryland, College Park, the Center for Engineering Concepts Development in the Department of Mechanical Engineering has developed a new course called *Engineering for Social Change*. Our new mantra is that "Engineering is not just Engineering", referring to the fact that much engineering work requires a significant percentage of soft skills, rather than simply technical skills. We also make the point that many types of difficult, global social problems cannot be solved by utilizing pure technical skills and require a new approach for success.

In our course, students learn about a variety of areas of current interest in which Engineering creates social change, giving them a better perspective on their role in the world. Topics have included the future of engineering, crowdsourced design, autonomy, robotics, climate change, energy entrepreneurship, and more. A unique component to our course is the integration of a philanthropic gift in an engineering class, deliberated and decided by the students. Courtesy of the Neilom Foundation,[431] a Maryland nonprofit working to improve the lives of young people, a $10,000 in grant funds is provided each semester specifically for careful and considered investment by the students in a local nonprofit organization creating effective social change. The students analyze the organization and its project proposal in an engineering area and vote as a class for the proposal that is considered to have the highest probability of a sustained positive impact.

The class is undertaken in conjunction with faculty from the School of Public Policy, in recognition of the fact that social change is often the result of major policy implementations and that most engineering advancements are regulated through various forms of public policy. Professor Robert Grimm and Jennifer Littlefield of the Center for Philanthropy and Nonprofit Leadership (CPNL) provide support in sharing the unusual idea of philanthropy with our students, then help to guide them through the process of selecting the nonprofit of their choice.

As an example of the type of content within our course, Mechanical Engineering Professor Jungho Kim lectures on the future of engineering, examining futurist Ray Kurzweil's predictions about the future of artificial intelligence and making projections as to what computing power may be capable of in the future. The past and future impact of automation on society is considered and discussed as a societal issue. The growing need for engineering to improve our world is discussed as society grows in numbers and needs. The National Academy of Engineering's "Engineer of 2020" is discussed, along with the variety of skills required to be the engineer of the future. The rate of technological change is detailed with its associated effects on society. Future technologies are presented, and multiple perspectives on societal impact of these are presented. The concept of global equality is challenged, and perspectives are given on the financial scope of humanity.

Michael Pecht, Chaired Mechanical Engineering Professor and Professor of Applied Mathematics, lectures on the impact of society on technology and vice versa. He compares how the culture in different countries has an impact on the development of even our most basic and fundamental technologies. He cites the toilet as an example, comparing versions from different countries and periods to explain cultural preferences in product design. The ideas of communications improvements, social interaction, salmon farming, genome editing, autonomous vehicles and military drones, brain-controlled computers, gender selection in babies and the "truth" of digital imagery are also examined with an eye on how engineering, technology and society interplay.

Of all the Engineering courses taken as requirements for an engineering degree, only a fraction are used by engineers during their lifetime in the practice of their profession. The curriculum, however, is designed to create an engineering mindset and a way of problem solving. It is for this reason that we want to inculcate an appreciation of the social change engineering creates and how not only for-profit enterprises but also philanthropy and non-profits act as catalysts. As students in our classes over two semesters have noted with some vigor, Engineering for Social Change should be a fundamental course and a requirement for all engineers!

References

[409] Dylan Hazelwood, "Bennett Hazelwood in Virtual Reality", Image, 2016.
[410] Alan P. Santos/DC Sports Box, "ENME 467 Grant Celebration", Center for Engineering Concepts Development, December 17, 2015.
[411] Attributed to George Herbert Allen, American Football Coach, April 29, 1918 – December 31, 1990.
[412] "NAE Grand Challenges", National Academy of Engineering, 2016, accessed at http://www.engineeringchallenges.org/challenges.aspx
[413] "Extreme", Stanford University, 2016, accessed at http://extreme.stanford.edu/
[414] Kelleher, K., "Why Intel is Laying Off Thousands of Workers", *Time*, April 22, 2016, accessed at http://time.com/4304632/intel-layoffs-pc-microsoft/
[415] "Engineering Tasks for the New Century: Japanese and U.S. Perspectives", National Research Council, p.64, 1999, accessed at http://www.nap.edu/read/9624/chapter/8
[416] Lupton, R., *Toxic Charity: How Churches and Charities Hurt Those They Help, And How to Reverse It*, HarperOne, October 2012.
[417] Zuger, A., "Folding Saris to Filter Cholera-Contaminated Water", *The New York Times*, September 26, 2011, accessed at http://www.nytimes.com/2011/09/27/health/27sari.html
[418] "Infusing Ethics into the Development of Engineers", National Academy of Engineering, February 18, 2016, accessed at https://www.nae.edu/150176.aspx
[419] "Code of Ethics for Engineers", American Society of Mechanical Engineers, February 2012, accessed at https://www.asme.org/getmedia/9EB36017-FA98-477E-8A73-77B04B36D410/P157_Ethics.aspx
[420] "NSPE Code of Ethics for Engineers", National Society of Professional Engineers, 2016, accessed at http://www.nspe.org/resources/ethics/code-ethics
[421] "Code of Ethics", American Society of Civil Engineers, 2015, accessed at http://www.asce.org/code-of-ethics/
[422] "IEEE Code of Ethics", Institute of Electrical and Electronics Engineers, 2016, accessed at http://www.ieee.org/about/corporate/governance/p7-8.html
[423] Atiyeh, C., "GM Ignition-Switch Engineer Speaks after Months of Silence: "I Did My Job"", *Car and Driver*, November 17, 2014, accessed at http://blog.caranddriver.com/gm-ignition-switch-engineer-speaks-after-months-of-silence-i-did-my-job/
[424] Ewing, J., "The Engineering of Volkswagen's Aggressive Ambition", *The New York Times*, December 13, 2015, accessed at http://www.nytimes.com/2015/12/14/business/the-engineering-of-volkswagens-aggressive-ambition.html
[425] "Mitsubishi Motors' fuel economy scandal spreads, affects 20 models", *Japan Times*, June 17, 2016, accessed at http://www.japantimes.co.jp/news/2016/06/17/business/corporate-business/mitsubishi-motors-fuel-economy-scandal-spreads-affects-20-models/

[426] "Mitsubishi Motors: How did it falsify its fuel economy data?", BBC, April 21, 2016, accessed at http://www.bbc.com/news/business-36099044

[427] Jacobson, L., "Statement on the Gulf Oil Spill and Licensed Professional Engineers", National Society of Professional Engineers, 2016, accessed at https://www.nspe.org/membership/about-nspe/statement-gulf-oil-spill-and-licensed-professional-engineers

[428] Cohen, S., Grace D., "Engineers and Social Responsibility: An Obligation to Do Good", *Institute of Electrical and Electronics Engineers Technology and Society Magazine*, Volume 13, Issue 3, Fall 1994.

[429] Baillie, C., *Engineers within a Local and Global Society (Synthesis Lectures on Engineering, Technology, and Society)*, Morgan and Claypool, 2006.

[430] Gates, W.H., "Engineers without Borders", Bill and Melinda Gates Foundation, March 8, 2008, accessed at http://www.gatesfoundation.org/media-center/speeches/2008/03/william-h-gates-sr-engineers-without-borders

[431] The Neilom Foundation, 2016, accessed at http://www.neilom.org

Fall 2015 Engineering for Social Change Student Project Winners [432]

Social Entrepreneurship at the College of Southern Maryland [433]

Appendix A

Engineering for Social Change: The What, Why and Who

A course introduction from Professor Emeritus Davinder K. Anand

Welcome!

This is a unique course in the history of our College of Engineering and I say that with much experience in this Department. I want to walk you through my thought process of why we have introduced this course and how we are going to work together this semester.

I am a professor, former Chair, have taught almost every offered course in this Department, led $40M+ of research, and published many papers and books. My areas of interest over the years were in satellite dynamics, alternative energy, manufacturing, system simulation, and VR. I have also worked in industry in the A/C field for the Carrier Corporation. I went to Syracuse and was trained at their factory. I am one of the very few registered PEs in our Department. I was a Program Director at NSF, I have been on the senior staff of APL/JHU, consulted for NASA, DOE, the Navy, the Army, and industry, and had my own consulting firm which I sold for a million bucks when I became Chair in 1991. Today, I have my own private foundation, the Neilom Foundation.

At the age of almost 76, after all this, I was not and am not content. Why? I don't have a precise answer for that. But I believe it has to do with the fact that I felt, and feel today, that what I learned as an engineer was incomplete. There was something missing. The engineering you are learning is basic and absolutely necessary and good. But we must, as part of our engineering education, connect to the social fabric of life!

In many ways the roles that engineers take on have always extended beyond the realm of knowledge and technology they learn. In fact, engineering impacts the health and vitality of a nation as no other profession does.

I know the formal definitions, as do you -- creativity, inventiveness, doing good for mankind and so on. I want you to read on Canvas about what the National Academy of Engineering thinks about us and where we are headed in terms of the curriculum that we should be exposed to. You can also find out what ABET says.

Today I ask the question: *Are we teaching everything that you need?* In other words, *is all engineering engineering,* and *what is it that all of you should know and be exposed to in order to be leaders, entrepreneurs, shakers and movers and even philanthropists?* This is important because one third to half of you will never perform engineering work. That is why I say *engineering is not just engineering*.

This class has about 40 students out of perhaps 500 seniors and juniors. That is 8%. So all of you are very special and are looking for something more than just traditional engineering. Why did you sign up for this course? I hope not to get a handle on some more theories or equations. Maybe you are thinking about leadership, or about becoming a global engineer. Maybe you are someone who wants to make a real difference. It is the kind of hunger that I believe some of our good students have.

So I go back to my own career, which is similar to that of many people whom I know who have been successful in engineering. The further up the ladder I went, the less engineering of the type you are learning I used!

Interestingly, even at the highest level as engineers we tend to shy away from taking credit for the great social change we bring about, the philanthropic things we do, and the fact we have so little say in public policy. So my discontent continues, which is amplified by other unrelated events.

Last year I accidently met this fellow in the School of Public Policy with the title of Professor of Philanthropy (actually Dylan Hazelwood found him). His name is Dr. Robert Grimm. He invited me to their Do Good Challenge, a campus competition to create social change. I was there for three hours! It just blew my mind.

We met several more times since then, and this course is the result of that first encounter. We put together an entire team and have spent many, many hours talking about what the content of this course should be. We argued about the name and team members, met once or twice a week, and on and on. Trust me; I have never spent so much time starting a new course in my 50 years on this campus. Many years ago, I believe in 1969, I started ENME 403: Introduction to Control Systems – I hardly spent five minutes talking to the Chair, and bingo! But not this course. Once the team was formed, we knew what each one of us was going to do. But we were not quite sure what you, our students, should get out of it. Although we have come to some broad conclusions, some of it is work in progress. I am going to request your help in determining that since this is a course for you and in many ways by you.

I think this is a good time to introduce my team, which consists of Mr. Dylan Hazelwood, Dr. Mukes Kapilashrami, Partners in Public Policy, Speakers, you (the students), and me. I want to say a few words especially about Dylan. He found the Professor of Philanthropy and has been most involved with me in selecting content and organizing every detail of this course. And he will tell you during the semester how difficult I have been in getting this thing off the ground. Never has this Department been so involved in starting a new course as this one. And you will appreciate that, I hope, as we go on.

I am reminded of a book about the *Rules of the Red Rubber Ball* by Kevin Carroll. Anyone know about it? There are two central themes as I see it. The first has to do with recalling when you were young. I mean very young -- 10 or 12, and you had a hobby or something you liked. You spent enormous amounts of time on it. Happily, willingly! Your mom said, "Stop that nonsense and come and have dinner!" But you kept at it. Why? Because it was a game you liked. You were totally engaged because it was a game. When we play we have fun. Time becomes irrelevant.

The second theme was to pass the ball. You remember playing soccer, basketball or another sport and one kid hogged the ball. Everyone is screaming, "Pass the ball for goodness sake!" But there is always this one kid, sometimes good and sometimes not, who just does not pass. It used to tick me off big time. I think I learned the art of patience at that time in my life. So I suggest you consider this unique course as a sport and that you pass the ball, meaning let everyone speak, and share their views and experience.

So what is expected of you in this class? Listen, question assumptions, and be engaged. Be passionate about an idea where engineering or technology plays a significant role in society and you can make a statement. Learn the needs of others where technology can help but you have not previously considered how it could help. See how philanthropy and the non-profit sector of our economy can be more usefully used to advance the idea that engineering can have a positive change for a large swath of humanity. At the very least, you will get a broad education!

I originally thought that I could come and talk to you about many projects and ideas of mine that have been dear to me for more than fifty years. I actually thought about doing just that. Then I vetoed my own idea and decided to bring in a newer and mostly younger breed of experts to talk to you, and then discuss with you critically what they have said. I didn't want you to just hear a talk and walk away. And so after each lecture we shall discuss what we have learned. Or sometimes the speaker and their talk might be a bust! You should be prepared for this to happen. So I have personally selected all the speakers, working very closely with Dylan. We have picked topics of interest that are also concerns of the day.

For example, I have asked Dr. Robert Grimm to talk about philanthropy. He is our partner from the School of Public Policy and runs a successful program in his Center. My good friend Professor Michael Pecht has traveled the world, and has an intimate idea of how electronics have made major impacts in our lives. We just brought into our Department Professor Jelena Srebric, who is passionate about sustainability. She has a very active and dedicated group supported by the National Science Foundation and works with industry and the government. Professor Jungho Kim in our Department has been active in Engineers Without Borders, and will talk about future trends in engineering. Dr. Mark Fuge will talk about crowdsourcing and big data. He is the newest faculty in our Department and is an expert on issues relating to the best use of big data. Dr. Jennifer Littlefield, Associate Director of the Center for Philanthropy and Non-Profit Leadership, will lead you through the process of investing the class's $10,000 in grant funds. I have an advisor to my Center who is a retired two-star admiral whom I asked to talk about Autonomy. Remember that less than 6% of the officers become admirals -- similar numbers as this class! We have also invited Suchita Guntakatta from the Gates Foundation to discuss vaccine delivery in Africa.

All of these folks satisfy my criteria of why they should be speaking to you – they are experts who have passion and/or expertise and use their skills to make an impact on society with their own ideas.

When I say *engineering is not just engineering*, what do I mean? Not only is the *curriculum* changing but so is the *environment*, each feeding the other. But the curriculum is not keeping up, and perhaps it cannot. In

1959, a BS took 144 credits and included courses like Kinematics, Introduction to Engineering and Use of a Slide Rule, Power Plants, Drawing, Electric Motors, and Advanced Thermodynamics -- industry-driven skills. Ductulators and slide rules were nomograms of the day.

By 1965, when I came to UMD, the number of credits required for a BS was reduced to 139. I think we eliminated Electric Motors and reduced Physics. When I became Chairman of the Department in 1991, we reduced it to 124 credits, which is where it stands today. It became consistent with any other BS degree on campus. So, was the engineer with 144 credits smarter than the one with 124 today? The answer is absolutely not! Many of the things previously taught have crept into high school. Also, with the introduction of technology (calculators, computers, laptops), it became easier to access information. We have also dropped some of the skill-based courses. This is consistent with the belief that you come to the University for knowledge and a broad education, not just to learn specific skills.

But something else has also happened in the field of engineering. Driven primarily by technology and new discoveries, engineering has become very broad and complex indeed.

We have activities in our Department that you have never heard of. When I first hired people in MEMS, Robotics and CFD, I heard a lot of grumbling from the old timers! We are very much like the medical field. With a large number of specialties, only a very basic engineering education at the undergraduate level is necessary. Also, this education has to recognize the reach of the engineering profession of today.

A large number of students today are not content. The bright students want more and question the very basic assumptions of life, and that is good. I remember seeing from my office window in the late '60s students blocking Route 1. National Guard soldiers with bayonets were standing guard in case the students misbehaved, since they were openly smoking, and not just ordinary cigarettes! There was an unease and discontentment in the air, but to be sure there was excitement also, and the students and even some faculty were determined to do something. The issue of the day was the Vietnam War.

Fifty years later, the same unease has returned. It is not only the issues in Ferguson and New York and Baltimore, but also our national politics, international conflicts, and social media. Information is broadcast almost in real time and all of a sudden, everyone is in on it.

This is all driven by technology, and that on the one hand creates unease, but on the other excitement. We all ask, "What can we do?"

I want to make a few remarks about engineering and social change using the cell phone as an example. Consider what has happened to the world with the introduction of social media. It is a wonderful thing bringing people together separated by miles. As a matter of fact it has made the world smaller (and indeed flatter, according to Thomas L. Friedman). But then it has also helped us lose touch with each other. We send an email or a text if someone is sick instead of visiting them in the hospital – the doctor has to take the bandages from my eyes so I can read all those get well messages. We send an e-card for a birthday, for Christmas and so on. We are connected to the world virtually and yet we are alone. I think about this feeling of isolation, instant messaging, virtual activities, automated answering… "Hello this is Linda and I am your virtual advisor." No, I want a human! And I slam the phone down. There are old people going bananas because they have no one to talk to! These are social changes that have been accelerated by technology and engineering – some good and some bad.

How can we differentiate, learn to use only some of the good stuff, and teach it to our students?

In 1967 or so, I used to discuss *The Tragedy of the Commons* all the time. It is an economic theory by Garrett Hardin, which states that individuals acting independently and rationally according to their own self-interest behave contrary to the best interests of the whole group by depleting some common resource. Land, the stock market, and now the Swiss Franc are excellent examples. Technology has played a key role in each of them.

It is not enough to spend eight hours a day and do the same old, same old. We want to make a difference. The new engineer is not the same, and engineering is not only engineering. These are big ideas. But in the short term what will you get out of this course? You will acquire the knowledge and mindset to pursue projects that make a difference. Besides listening and learning and discussing, you will also do something unusual. You will join your class group in discussing, selecting and presenting a $10,000 award (sponsored by Neilom Foundation) to a deserving non-profit.

There are issues that will be raised in the course that I do not have answers for yet. Some of the issues you will have individual views on, others we all might agree upon. For example, what metrics would you use to describe an act of engineering and what metrics would define the minimum unit of social change you're thinking about? And what timelines are you thinking about for the engineering and social change topics? Is it years, decades, or generations that you're considering? Or perhaps it's just the length of a student's professional career? And how direct is the path you're considering between engineering and social change? Is it very direct or convoluted, as in much of life?

Let's say you improve the yield of food production (via agricultural engineering or genetic engineering) to allow for more personal free time and more wealth within the farming population. Does this lead to more education, better jobs, smaller family size, larger farms, displaced workers, need for hired migrant workers, depleted soils, loss of natural habitat and water resources, more congested cities, more crime, etc.? What is good and what is bad?

Social change will happen no matter what. So are we talking about good engineering and good social change – as opposed, for example, to unsustainable technologies (i.e. good in the short run but disastrous in the long run, like oil or coal) and bad social change?

Can we even hope to plan or even predict the route between engineering and social change, or will it always be more or less random in nature depending upon the larger political and egocentric power forces of the world? And how much of engineering is based upon the accidents of collaboration and of what's happening in disparate technical fields and/or resource discovery and availability?

And what about good technology versus bad technology? Or is technology value-neutral, in which case it is the use of that technology by man that gives it positivity or negativity? In what kinds of cultural (social) environments does the best engineering happen? Is that even a meaningful question?

A more direct route between engineering and social change may be entrepreneurial. Bill Gates is pushing social change in education and health improvements in third world countries. The success of this endeavor is under discussion!

It is provocative to think about who actually controls the path between engineering and social change. Politicians? Corporations who buy engineering patents to suppress or undercut competitors? Governments who pay for only certain types of research? Capitalists?

One of my friends just visited a third world country and stayed in a beautiful hotel and saw a woman cutting the grass with scissors! What do you make of that? I visited a third world country where I saw a thirty-plus floor building built using the most modern technology alongside no technology at all! Bricks were being carried and delivered by a donkey or on a person's head. The social implications are mind-boggling to say the least! My travels convinced me that much of useful technology has not socially impacted a large swath of humanity.

So as you move into the future, what should you expect?

The economy in which we will work will be strongly influenced by the global marketplace for engineering services, a growing need for interdisciplinary and system-based approaches, demands for customization, and an increasingly diverse talent pool. The steady integration of technology in our infrastructure and lives calls for more involvement by engineers in the setting of public policy and in participation in the civic arena.

At their core they call for us to educate engineers who are broadly educated, who see themselves as global citizens, who can be leaders in business and public service, and who are ethically grounded.

Let's talk about the attributes needed for the graduates of 2020. These include such traits as strong analytical skills, creativity, ingenuity, professionalism, and leadership.

The business competitiveness, military strength, health, and standard of living of a nation are intimately connected to engineering. And as technology becomes increasingly integrated into every facet of our lives, the convergence between engineering and public policy will also increase. This new level of intimacy necessitates that engineering and engineers develop a stronger sense of how technology and public policy interact.

The engineering profession recognizes that engineers need to work in teams, communicate with multiple audiences, and immerse themselves in public policy debates, and they will need to do so more effectively in the future. In the face of pressure, especially from state funding agencies to cut costs by reducing credit hours for the four-year degree, it remains an open question whether engineering education can step up to the challenge of providing a broader education to engineering graduates.

I guess my mental ramblings have led to the question I should have asked you first. What do I want you to take away from this course? The answer is, I do not know!

The formal statement for this course says that you will learn how engineering creates social change and how philanthropy and non-profit organizations are catalysts. I actually doubt it. But I do think it is fair to say that you will learn enough to start asking many questions at the intersection of engineering, philanthropy and social change and how they may, will or should impact public policy. These are big issues with no easy answers.

At the end of the day, I hope that this course and our conversations will create a legitimate subject of study and inquiry at the intersection of engineering, philanthropy and social change, and that you will enjoy starting something new in the College of Engineering and being part of this community.

Davinder K. Anand
January 26, 2015

Appendix B

The Students' Voice: Blog Postings

Spring 2015 Blog Post Excerpts

"This class has done more for me as an engineering student interested in social change, specifically through entrepreneurship, in all of my collegiate experience at this university....What did the most for me were the Monday lectures held by guest speakers." – D. Barotti

"In this course the class learned that there is a surprisingly strong connection between engineering and social change. Most engineers know that their work goes on to impact the world, but usually the full impact on society is not considered." – J. Batterden

"Writing an RFP as a class and talking with non-profits in a class selected area was wonderful. A completely new experience in dealing with people; as students we are used to begging for money and writing for scholarships. This time we were on the other end of the grant and it was a new perspective. This has been an eye opening class that all engineers should take." – N. Bellis

"Being able to participate in the grant making process and actually be able to donate $10,000 to a non-profit organization that would put our money towards helping the community was an incredible experience." – J. Chen

"Although the Engineering for Social Change course may not have been the most technical course in my schedule, there is no doubt that I was still able to learn a lot throughout the semester." – K. Chin

"...one thing that I believe that I took away from this class that will be essential to my life. This essential life lesson has not been taught in any formal engineering class in my academic career. The lesson is to 'not be afraid to be unique and strive to make the world a better place in your own way'. We are constantly taught to improve the world by making things more sustainable, smaller, faster and better looking. But we are not encouraged to be different, and change the world with non-engineering ways. But this course does just that!" – J. Clagget

"Without philanthropic intentions, the role of engineering in the world would be driven entirely by financial matters, and would not necessarily have as many positive outcomes for society." – M. Kafer

"When I first went to the class, honestly, I was little disappointed because the first thing they did was spreading the chair around and started discussing. I personally am not a person that talks much in class, so I did not like it much. I even had to look up the work "philanthropy" because I did not even know what it means. However, now that I think of it as at the end of the semester, the class Engineering for Social Change was a class that I would never forget in my life and have no regret taking this course." – D. Kim

"Dr. Anand informed the class about the existence of the Neilom Foundation and then told us that we would all be entrusted with $10,000 to invest in a non-profit company of our choosing. To make matters more intense, we then found out that as students, we could also score $1,000 for individual ideas to start our very own non-profit!" – A. Kusimo

"The course Engineering for Social Change may have originally been an experiment to see how students view the ties between the technical aspects of engineering and the social impacts of this technology. However, in the end it became a prime example of how difficult it is to get 30 individuals from vastly different backgrounds to agree on something as passion-inducing as awarding $10,000 to a nonprofit organization." – E. Love

"'Social change' is not a phrase that usually comes up in engineering classes. Engineers are supposed to care about numbers, programming, and making things faster, stronger, and better – right? Well after this semester I can definitively say engineers need to be considering much more." – S. Niezelski

"While studying engineering I think we often get wrapped up in the equations, the formulas, the applications, and how we can advance technology at any means. ESC made a human connection with engineering, and challenged us to consider how we affect society through engineering and vice versa." – J. Pinker

"An engineer has an ethical responsibility with the part of society that he or she influences. This responsibility, often overlooked, is essential in understanding how philanthropy ties into the profession of engineering."
– S. Rajaram

"The course showed me that with passion and determination, anybody can make a difference." – A. Savage

"Prior to matriculating into this course, I viewed engineering in a certain way. I primarily saw it as a way to be innovative and contribute to technology. Selfishly, I also saw it as a means to create and live a comfortable and luxurious lifestyle. Today, it holds a lot more value than innovation and income. Through this course, I have been able to connect engineering to matters I am very passionate about, world development and sustainability." – W. Sama

Fall 2015 Blog Post Excerpts

"As an engineer, the best answer or the most efficient solution does not always correlate to what is best for the community you are attempting to help." – T. Aion

"This class spent the majority of its time breaking down the fourth wall and attempting to reinstate peoples individual values back into their engineering toolbox." – B. Avadikian

"My repetitive use of "perspective", "eye-opening" and "realized" is why this class is important to engineering students." – J. Brio

"Engineering is all about solving problems, but most engineers don't think about what problem that they're trying to solve. Classes such as heat transfer, vibrations and electronics teach us how to solve math problems. In the real world, we're going to need to solve real problems in the world. This class taught me a great deal about the negative impacts of ignoring our impact on society." – J. Chen

"To be totally honest, I wasn't really sure what to expect when I walked into Engineering for social change for the first time, but I at least had hope that it would be different. I was not disappointed." – C. Clarke

"For much of my school career, I have stated that my greatest lessons in engineering have come outside of the classroom. I have always felt I learned more from my internships, and my acts in helping clubs or organizations around campus develop fund raising strategies. Engineering for Social Change has directed me away from those thoughts during my last semester here at UMD." – M. Froeschle

"Taking the Engineering for Social Change course has really reminded me why I chose to be an engineering student in the first place." – J. Hart

"The grant process was humbling, yet fulfilling." – N. Lapides

"My ISCC project concerns providing potable water to a select region in Ethiopia. There is so much that needs to be done here, but many other people do not have the education and knowledge to implement sustainable solutions. Previously, I have taken my knowledge for granted, but I now understand that I can help make a significant difference in people's lives." – K. Lee

"From the guest speakers and by working on the ISCC project with a team, I have learned that the most efficient, cutting edge, technical design is not always the best design for society. When creating products, engineers need to consider more than the functions of the product, but the wants, needs, and beliefs of the end users." – K. Parker

"When Dr. Shelby, from USAID, came in to give his guest lecture that focused on the co-design process, it was a really eye opening learning experience. It was not surprising to me how important it is to consider the needs of others, but how easy it is to simply overlook the fact that someone else may define their needs differently to how I may see them." – T. Salvador

"One must always consider values and ethics as an essential factor in any engineering decision making process." – L. Terry

"Engineering for Social Change: Those four words may have saved my impression of engineering. It was the idea of my work and struggles as an engineer could be used on a larger scale to directly impact others. It was the fact that communities of disenfranchised and impoverished people had been conveniently overlooked throughout the first three years of my coursework. It was the premise of engineering having a greater purpose than fighting to complete my weekly homework assignments. To me, it is simply an understatement to try and say how the class has influenced my opinion about engineering: it has preserved my sanity and found a new passion within engineering in which I can explore." – C. Urrutia

Student Feedback: In-Class

During the final lecture of each semester, Professor Davinder K. Anand (Course Leader) hosted a "round table" meeting with all of the students to not only gain their personal view and feedback on their journey through the Engineering for Social Change course, but more importantly, to understand their perspective on the various components of this course (as listed below) to further optimize it to best meet the need of the students:

- Engineering and Social Change Lectures
- Philanthropy Lectures
- Neilom Engineering for Social Change Grant
- The Ideas for Social Change Challenge (ISCC)
- Blog
- Grant Ceremony

Lightning Round Feedback

- "This class humanized my degree choice and my engineering education."
- "Cut down the number of lectures, and instead expand the discussion part."
- "In this course we learnt, for the first time, about the consequence of our action rather than just the outcome."
- On the topic of voting for the grant process – "we were given too many options".
- "The ability to be creative and make their own decisions in the class was a very good thing."
- "We need more emphasis on proposal writing and analysis – not enough time spent on this."
- "This was a good class, unlike any others in Engineering. It was good to get the students' opinion and helped shape my own mind."
- "This was a different kind of class. The best thing was to debate and come to a decision together as a class."
- "It was good that it was not just technical. Also it was hard to know if the decision we made was the right one."
- "I enjoyed the discussion aspect of this class, and also policy issues, which is not there in the typical engineering courses."
- "It was great to see so many millennials getting involved in social change."

- "I liked the guest speakers and the ISCC project – to work on a technical idea for social change. Different viewpoints and focus on people was good, not just the technical side."
- I liked looking at the difficulties in issues around solutions and not just the technical part.
- "I liked the discussion aspect of this class, and enjoyed hearing other opinions."
- "I liked the guest speakers, in particular the guest speaker from Seattle (Smeeta Hirani) – usually they talk about external issues but this was internal for her. I would like to see how more products affect social change, not just philanthropy."
- "I think the people who need this course the most won't sign up for it."
- "Engineers are problem solvers – this was the only class to question the personal definition of problems."
- "Our impact in the real world lies beyond the classroom – I'm always in a rush to normally get out of the classroom, but this was different."
- "I liked that we had real money to give away, real situation with high stakes. This was the only class where I felt the course is about more than my own education."
- "The discussions were good – I have been in class with the same students and have never heard them speak before. I am glad I took this class and hope it will be given in the future."
- "I think this course should be a requirement for all engineering students."
- "This class was the first time I had used my social conscience – it should be offered everywhere!"
- "I really liked that the course wasn't too technical, or stressful and unpleasant to come to – I think soft skills are important. Smaller group discussion would make things easier."
- "I liked the amount of perspective and empathy gained from this class, and it opened my eyes to the world – a far cry from mindless mathematical minutia. Have students be more involved in facilitating discussion."

Appendix C

Spring 2015 $10,000 Neilom Engineering for Social Change Grant: *FRESHFARM Markets*

At the culmination of the Spring 2015 offering of CECD's Engineering for Social Change course, Mechanical Engineering students chose FRESHFARM Markets, a nonprofit from Washington D.C., for the $10,000 class fund provided by the Neilom Foundation out of 15 proposals submitted. This will fund the building of a 55' by 55' edible garden space at Ludlow-Taylor Elementary school, a Title I public school in Washington D.C. serving many low-income students.

FoodPrints is FRESHFARM Markets' local foods school program that builds an edible schoolyard garden and integrates the garden into the school curriculum. FRESHFARM Markets is a nonprofit organization whose mission is to build and strengthen the local, sustainable food movement in the Chesapeake Bay watershed. They do this by operating producer-only farmers markets and innovative outreach and educational programs that educate the public about healthy food and related environmental issues. The students were impressed by the organization's history of success in DC schools, the level of integration of the project into the curriculum at and partnership with Ludlow-Taylor Elementary, careful project plan and budget, and the long-term impact the grant funds would have. Children who participate learn about where their food comes from and how important it is to eat fresh, nutritious, seasonal foods, a key consideration with rising obesity rates among young children.

Appendix D

Fall 2015 $10,000 Neilom Engineering for Social Change Grant: *Bread and Water for Africa*

At the culmination of the Fall 2015 offering of CECD's Engineering for Social Change course, Mechanical Engineering students chose Bread and Water for Africa, a nonprofit from Alexandria, VA, from the 10 proposals for the $10,000 class fund provided by the Neilom Foundation. This will go towards building a hand pump water well to serve the Hill Station Primary and Secondary Schools as well as the surrounding community in Freetown, Sierra Leone. The vast majority of Sierra Leone does not have access to safe and clean water and nearly half of the population uses unprotected water as their primary source for drinking, bathing and washing. Bread and Water for Africa's mission is to promote positive change in Africa by supporting and strengthening grassroots initiatives for community self-sufficiency, health and education. Through partnerships with grassroots organizations in Africa, they have provided water wells for tens of thousands of people in Kenya, Uganda, Ethiopia, Mozambique, Zambia and Sierra Leone. The students were impressed with the organization's experience and success in working with regional partners to accomplish past well projects, demonstrated success in executing plans for self-sufficient ventures, and recognized the strong need for a well in this community. It is estimated that the Freetown well will serve a community of 2500+ for the next 15-20 years, and will save many lives.

Appendix E

Engineering for Social Change Student Project Work

Spring 2015: Virtual Nonprofit Challenge (VNC)

The Virtual Nonprofit Challenge project asks student teams to develop an idea for a nonprofit initiative that would utilize both their engineering skills and mindset to support a well-defined community need. The project may be an entirely new initiative or an expansion of an existing real-world initiative, although if it is an expansion of an existing program students must keep in mind there must be significant evidence of innovative new thinking around the predefined problem. Student teams will develop their "virtual nonprofit," including their ideas as to how they will secure more funding beyond an initial virtual $1,000 startup grant. They will present their organization and plan to the class, who will vote for the winning team. The winning team will then be granted $1,000 by the Neilom Foundation to either turn their virtual idea into a reality, or to donate to an external nonprofit of their choice.

The Project

This will be a full semester project and will count for 25% of your final grade in this class. Students will be assigned in teams of five (5) and work together throughout the semester in these teams. Team project reports should be no more than ten (10) pages for distribution to the class, and they should answer the following kinds of questions:

- What are the values of your organization?
- What is your organization's mission statement?
- Who is the community served by your organization?
- What other existing organizations work in this space?
- What are the specific tasks the organization will undertake to accomplish its mission?
- What are the core competencies required for a nonprofit working in this area and how does the experience of each of the team members help to fulfill that need?
- How does engineering play a role?
- What is your detailed budget for year one and what is your fundraising plan?
- What are your proposed metrics for measuring efficacy?

The second will be a group presentation of no more than 20 minutes. It should include the following key points:

- Introduce your nonprofit (vision, mission) and detail the community issue that it addresses.
- Explain how the nonprofit will address that issue, how it will measure success, and how engineering plays a role.
- Illustrate how each team member's skill and experience helps fulfill the mission.
- Present a plan for growth, and convince the class this nonprofit is worthy of their support.

Students in the class will vote after each presentation, then the winning group will be given the opportunity by the Neilom Foundation to use $1,000 towards developing their nonprofit idea, or they may choose to donate the funds to a nonprofit group of their choice to support disadvantaged young people.

VNC Student Team Final Report Abstracts

Inspyre - The Interactive STEM education experience – Award Winner
(J. Claggett, M. Johnson, W. Sama, B. Brian)

The United States is losing its competitive edge in math and science while the rest of the world soars ahead. The way in which we fix this is by reaching students at a young age, supporting and encouraging interest in STEM fields. "Inspyre" would like to assist in this effort is by educating young Americans in the Science, Technology, Engineering and Mathematics by providing tutoring, workshops, activities and a plethora of other STEM education related services to communities in the local DC area. We will be serving elementary schools, in Prince George's County and D.C. public schools, specifically pre-kindergarten to 6th grade. We chose this target because at this age, students are in the beginning stages of their academic development. Teaching elementary school STEM subjects will help orient their minds from a young age towards STEM courses. Such familiarity at this tender age range will not only help the students learn the material but encourage them to potentially pursue careers in these fields. We noticed that students start off strong in STEM classes, and then lose interest as time goes by. Within D.C. schools, the 4th grade math proficiency level is 22%, whereas the 8th grade math proficiency level is 15%. Overall 46% of elementary students in DC are below the math proficiency average. This trend persists in national statistics. The average math proficiency level for 4th graders is 36%, while 8th graders have a 26% level (How are States Performing?). We aim to improve students' interests in STEM course, in that way their proficiency levels are increasing by grade level, rather than decreasing.

Team Human Energy
(D. Barotti, H. Desai, M. Hallenbeck, D. Kim, R.Portney, I. Zanko)

Our mission as a non-profit 501(c)(3) organization is to research, design and create wearable piezoelectric uniforms and equipment for those in the uniformed services. As we head further into the digital age, we grab our servicemen and servicewomen in electronics including radios, night vision goggles, and GPS systems just to name a few. According to Keith Johnson, staff reporter in the Washington bureau at the Wall Street Journal, this equipment and accompanying batteries can add an extra 16 pounds of weight to be carried by already heavily burdened soldiers. Our non-profit will work to create a uniform that can harvest the excess kinetic energy from the wearer's movements to serve as the energy supply for their electronic devices using piezoelectric fiber technology. In this way, our troops are their own sources of electricity, and do not have to carry around the extra weight of separate batteries, and will never worry about being out of charge. We want our organization and products reflect the values of those that would benefit from them. We are committed to designing and producing human energy harvesting gear as resilient, and effective as our servicemen and servicewomen, through research and development. Since our product is still in its developmental phase, we value creativity and innovation in design intensely to be able to provide the highest quality uniform with the maximum power yield.

Musical Minds
(K. Chin, B. Gotwalt, K. Herman, Z. Hutcheson, S. Rajaram, P. Yuan)

We at Musical Minds believe that all people have the right to an enriching, rewarding childhood and life. Our focus is on orphaned and disadvantaged children in group homes. We strive to educate and instill in these children a love of life, filled with wonder and the knowledge that they are empowered to create their own happiness. This is reflected through the programs we create, educating children in music and the culinary arts, giving them skills that they can use for both practical and enjoyment purposes.

<u>Serving Baltimore</u>
(J. Pinker, N. Bellis, J. Chen, J. Bendinelli, J. Batterden)

Serving Baltimore's mission is to bring nutritious meals to those who cannot afford it in Baltimore City. There are about 150,000 people living below the poverty line, which are our primary beneficiaries. Outside of this population, there are those who also cannot secure healthy food on a regular basis due to a lack of infrastructure. Baltimore City has several food deserts - defined as areas that lack a source of healthy food within walking distance. By providing an influx of nutritious meals to food pantries in these areas, Serving Baltimore will help make them more hospitable to their residents. The second pillar of Serving Baltimore is to strengthen the sense of community and pride of inner-city residents. Baltimore is a large city with a deep culture, and by combining this Baltimore spirit with the philanthropic spirit, Serving Baltimore seeks to infuse the power of doing good into the community. To do this, Serving Baltimore will highlight local artists and community activists that already serve the community, and give them a platform to spread their message. This way, Serving Baltimore becomes a place that strengthens the body with healthy food, strengthens the mind with engaging content, and strengthens the soul with a sense of belonging to a compassionate society. We believe this synergistic approach is the best way to serve those struggling to live in inner city Baltimore.

<u>E-Vest</u>
(N. Fikru, S. Hearne, A. Kusimo, E. Love, L. Rogovin)

E-Vest is dedicated to sustainably promoting Science, Technology, Engineering, and Mathematics (STEM) education in underprivileged low-income middle schools in Prince George's County by building a partnership between qualified undergraduate students from the University of Maryland and involved members of these communities. These partnerships will serve to excite students with the prospects of STEM education through semester-long multidisciplinary projects designed to introduce STEM related fields and concepts. E-Vest will work hand-in-hand with pre-existing Prince George's County middle school organizations to help promote STEM education and equip students with the skills they need to lead successful future STEM careers.

Pathways to success - Instilling Today's Youth with Goals and the Tools to Achieve Them
(M. Kafer, S. Niezelski, R. Rubin, A. Savage, J. Wiley)

Pathways to Success is a nonprofit that offers short term classes to rising 9th graders in order to help them gain perspective and prepare for a productive high school career starting in the fall. The organization will provide students with a broad picture of their options for post college life. Not only is a traditional college a wonderful option, but there are many other alternatives that are a better fit for certain people. Thus, Pathways to Success will not only highlight certain paths, but rather give the students some exposure to all of the different options they could begin working on next year. By practicing what the organization teaches, Pathways to Success will show and tell its students about the best practices and positive habits to start early.

Fall 2015: Ideas for Social Change Challenge (ISCC)

Objective

The *Ideas for Social Change Challenge* (ISCC) offers the students an opportunity to undertake an innovative and entrepreneurial approach to identify a problem (within the frame of a pre-defined topic) either in our local community or globally, and asks them to design a sustainable, holistic solution using their engineering skills. The students will work closely together in groups of 6-7 members on their projects throughout the semester, both in class and outside class hours.

The final outcome should emphasize the soft skills of an engineer and reflect on following 10 key requirements (out of which 7 need to be satisfied):

- Simplicity
- Low cost
- Use of local materials and resources
- Ease of transport and assembly
- Low maintenance and high durability
- Efficiency and functionality given the target community
- Environmentally friendly
- Social relevance
- Safety
- Innovative

The approach the students are expected to take is to create a high quality, comprehensive solution that reflects their deep analysis and understanding of the identified problem and the community in which the solution would be adopted. Further, students must propose a solution that not only satisfies the requirements listed above, but that has also been designed to be sustainable over the long-term. Many efforts to solve technical and social problems, particularly in developing countries, have been successful initially but without proper planning, they fail over time. Please note, the task is not to design a device but an overall solution for how technology/engineering can be implemented to solve an identified problem given its surrounding infrastructure and the end user(s).

The projects will be graded based on the students' performance throughout the semester (presentation in class and submitted reports). In addition to a written report, each group will present its project at the grant ceremony at the end of the course where the best project will be voted upon by audience choice (which will include faculty members, students and external guests) based on the criteria listed above. Upon completion of the ISCC, the students will have gained knowledge, maturity and experience in:

- Identifying a social problem and related concerns
- Connecting their engineering skills to a real world problem
- Developing empathy for end user(s) for better requirements-gathering and to develop a more appropriate and sustainable solution
- Designing a holistic solution to a social problem
- Public speaking

ISCC topics:

- Water crisis – *access to water, access to potable water*
- Renewable energy – *solar power, wind power*
- Food crisis – *food accessibility, power irrigation*

Student Team Final Report Abstracts

<u>Food Accessibility in Baltimore City</u> (Fall 2015 Award Winner)
(J. Schueler, J. Lofton, J. Chen, J. Hart, P. Thang, C. Clarke, W. Falak)

Food accessibility is a problem that takes many shapes and forms. A prominent issue related to food accessibility is the existence of food deserts. According to the United States Department of Agriculture, "food deserts are defined as urban neighborhoods and rural towns without ready access to fresh, healthy, and affordable food" (Food Deserts). Some crucial factors to consider when discussing food deserts are malnutrition and obesity; since fresh, healthy, and affordable foods are unavailable, residents are forced to turn to foods that are cheap but often unhealthy. These choices lead to vitamin and nutrient deficiencies and often obesity. Previous efforts to address this problem include food recovery, grocery route transportation, and urban agriculture. In this report, the approach to solving food deserts in Baltimore City will be to develop a rain barrel-fed aquaponic garden to provide fresh produce and fish. This aquaponic system is innovative due to its minimal water usage and low maintenance, and it attempts to combat the nutrition issue by providing healthy food at a low cost to local residents. Not only does this community garden help solve problems related to food accessibility, it provides an opportunity for the community to engage in cultivating their own food.

Implementing Wind Energy on Pine Ridge Indian Reservation
(M. Corboz, M. Gotsch, E. Herrera, M. Magnusen, K. Parker, D. Tarr)

The Pine Ridge Reservation located in South Dakota is among the poorest areas in the United States boasting an appalling unemployment rate of over 80% and an average annual income per capita ranging from around $3,000 to $4,000. Evidently, a large portion of this small income is directed towards running water and electricity. As a Native American community, the Oglala Sioux of Pine Ridge are incredibly environmentally conscious. Between economic troubles and environmental friendliness, the Sioux population would benefit on several fronts.

A wind turbine project, funded primarily through a US Department of Agriculture Rural Utilities Service (RUS) loan along with a 50% match from the US Department of Energy (DOE), could boost the Pine Ridge economy through installation and maintenance jobs as well as more affordable electric services while satisfying philosophical or indigenous beliefs and values with respect to the environment. In doing so, this project can truly transform a large community and have serious social impact. Previous projects including the Rosebud Sioux reservation project have created an incredible precedent for future projects with regards to funding and economic benefits. This project, located in a windy region of South Dakota, fits all criteria for successful implementation in the Pine Ridge region.

Access to Potable Water
(B. Avadikian, T. Cook, M. Damon, N. Lapides, K. Lee, F. Martino, C. Urrutia)

Our team explored the implementation of a lowtech water filtration device for use in Ethiopia, in parallel with an educational campaign on sanitation and cleanliness. The specifics of the device will be developed using data collected directly from the community's water collecting habits. Our team assessed data collected by the world health organization [1] to find that Ethiopia ranked second on the list of countries with the least sustainable access to improved/clean water sources. In order to narrow our scope down to our target community/types of communities, the team will only consider areas that have access to a water source where the water is not safe for drinking. In addition to our technical approach to selecting a region internationally, the team also considered the political and diplomatic relationships of the locations as well. In our case, the United States has a preexisting relationship with Ethiopia where "the United States and Ethiopia work together to enhance food security, improve health services, strengthen education, promote trade, and expand development." [2] With the backing of the international community, our efforts to secure clean drinking water to these areas can create positive social change via system replication and continued water sanitation education.

Solar Powered Drip Irrigation in Francisco Morazan, Honduras
(K. Villatoro, M. Liu, A. Patel, S. McDermott, J. Biro, L. Terry, E. Weiss, J. Argenal)

The people of Francisco Morazán, Honduras are suffering enormously from malnutrition and food insecurity recently due to severe droughts caused by El Nino Southern Oscillation (ENSO). With an economy focused on agriculture, it is critical to help the Honduran people grow their own food. In order to do this, we researched Honduran culture and current agricultural methods. The research showed locals are willing to do the heavy lifting, but we need to avoid the local government and instead partner with organizations such as SELF and the USAID. We isolated outdated irrigation techniques as an area to address in order to improve food security. Taking advantage of the dry conditions, we chose to set up a solar powered drip irrigation system that would run a maize or banana plantation. Experts help set up the SPDI system using local materials and teach the locals how to maintain it and make new ones along the way. The locals then run the SPDI powered plantation and create a sustainable economy with the system. In the short term, the

SPDI would reduce the rate of malnutrition. In the long term, food insecurity will no longer exist in Francisco, Morazán.

Access to Water in Syrian Refugee Camps
(C. Barnes, L. Choudhary-Smith, A. Hubbard, R. McGuire, M. Shin, V. Wu)

Syria's Civil War has caused large amounts of turmoil for its citizens. To keep their own families safe, many Syrian citizens risk it all to emigrate to neighboring countries and continents. Lebanon specifically has seen the largest influx of Syrian refugees compared to its pre-Syrian Civil War population such that one in five people in Lebanon is a Syrian refugee. This large influx without government support let the Syrians roam to do what they feel necessary, which leads to a number of problems within Lebanon. The most notable issue lies within the lack of access to water for Syrian refugees. Our solution is to take advantage of the rainy season by using a fabric that hangs from the roof of already designed refugee tents, acting as a funnel, to direct water into a bucket. The fabric will be supported at the edges to make sure water doesn't spill over in the lateral direction. Such a fabric will allow easy packaging and portability. We plan on measuring our success of the solution by checking for any decrease of crime or dehydration reports, maintaining the understanding that these aren't solely related to water access. Additionally, by improving the quality of lives of the refugees, we hope to measure the general health levels of the refugees, as we would expect fewer diseases to present. While we are not trying to solve the root cause of the refugees being displaced, we do hope to improve their quality of lives until they can return home.

Engineering to Bring Affordable Energy to Low Income Communities
(T. Aion, N. Carriere, F. Cova, Z. Frey, M. Froeschle, K. Rugel, T. Salvador)

In the middle of a heated political climate, debating over the growing income inequality, increasing energy demands, coupled with the rapidly heating global climate, it has become increasingly important to facilitate access to affordable clean energy. As solar technologies increase in efficiency and prices decrease each year, it becomes a more viable option for clean energy. When used in conjunction with government initiatives, it could also help to simultaneously strengthen low income areas. This paper will explore the potential economic, environmental, and social impact of a pilot effort to install solar panels throughout the community of Deanwood, in Washington, D.C., which has a proud history of embracing supportive change and new culture. This initiative will be made possible by the Department of Housing and Community Development (DHCD) who have already identified this community for a "Cutting Edge Affordable Housing Project Focused on Sustainability," Through the establishment of a local presence, with an emphasis on a sustainable solar solution that provides an easier standard of living to energy efficient homes, our pilot aims to incentivize a transition to a greener philosophy to be shared among Deanwood and neighboring residents.

Appendix F

Grant Ceremony and Celebration

Students and faculty in our Spring 2015 pilot course [434]

At the end of each semester the students attended a grant ceremony and celebration, held on the same day as the scheduled day for a final exam for the class. This event was comprised of the following components:

- An overview of the course presented by Professor Davinder K. Anand (Course Leader) followed by highlights and "lessons learned" from the semester, including the challenges experienced from an instructor point of view presented by Mr. Dylan Hazelwood (Course Manager).
- Student presentation of their semester long project work (Fall 2015).
- Panel discussion where students reflected on their own experience from the course as well as their view on the importance of incorporating social awareness in the curriculum of engineering education emphasizing the outcome of engineering (Spring 2015).
- Presentation of the $10,000 Neilom Engineering for Social Change Grant to winning non-profit organization.
- Reception and poster session where students presented their VNC projects to reception guests.

In addition to the students, staff and faculty members from Department of Mechanical Engineering and School of Public Policy, local community members from the State of Maryland and District of Columbia, as well as students from across campus, family members of presenting students, and local non-profit representatives attended the event.

Class of Spring 2015

At the conclusion of the Spring 2015 pilot semester for ENME 467, student team Inspyre (*The Interactive STEM education experience*) was the winner of the semester-long virtual non-profit challenge, as unanimously voted by the students in the class, and was awarded $1,000 by the Neilom Foundation to establish an operating non-profit organization to serve elementary schools in Prince George's County and D.C. public schools, specifically pre-kindergarten to 6th grade with a focus on STEM education.

The Spring 2015 theme of Neilom Engineering for Social Change Grant was access to local healthy food. A total of fifteen non-profit organizations working around the DC metro area responded to the request for proposals sent out by the students for the $10,000 Neilom Engineering for Social Change Grant. Following careful evaluation of all the proposals, including phone interviews with nonprofit leadership and on-site visits to the four finalist organizations, the students (in the capacity of being the grantmaker) came together and supported the work of FRESHFARM Markets, a nonprofit from Washington D.C. The students were impressed by the organization's integration of garden spaces, healthy food planting and harvesting, and food education into the curriculum in their FoodPrints program in multiple Title I elementary schools.

Class of Fall 2015

As a variation from the Spring 2015 session, the best student project in the Fall session was selected by audience vote at the grant ceremony following a five minute presentation by each student group. Each team did their best to highlight their novel approach to solve a social problem they had identified within a particular topical area among *water crisis*, *renewable energy* and *food crisis* topics. Although all student teams performed very well, and presented well thought out and innovative solutions highlighting both their engineering and social entrepreneurship skills, Team Food Accessibility in Baltimore City was announced as the best student team project.

The Fall 2015 theme of the Neilom Engineering for Social Change Grant was access to water. During this semester, ten nonprofit organizations submitted proposals in response to the request for proposals sent out by the students for the $10,000 Neilom Engineering for Social Change Grant. After careful evaluation of all the proposals by the students, again including phone interviews and on-site visits in the capacity of being the grantmaker, the students came together and supported the work of Bread and Water for Africa of Alexandria, VA. The organization proposed to match $5,000 to the $10,000 grant funds to build a water well in Sierra Leone, Africa, a region hit hard by the Ebola crisis in recent years and suffering from a serious lack of access to clean water. The water well would support 2,500 people for the next 15-20 years, saving lives and reducing illness.

Appendix G

Syllabus – Engineering for Social Change ENME 467 Fall 2015

Catalog Description

Critical analysis of issues at the intersection of engineering, philanthropy and social change. How engineering design, products and processes have impacted social change in the past and will do so in the future. Topics covered include energy, sustainability and climate change, autonomy, the digital future, low cost engineering, manufacturing, ethics and the impact of electronics on society. Faculty and external experts will engage with students on these and other topics of current interest in engineering and social change. Students will award a significant amount of grant money to an organization involved in technology for social change.

Course Vision

To inculcate an appreciation of the social change that Engineering creates, and how not only non-profit organizations and philanthropy but also for-profit enterprises act as catalysts. Students will appreciate that there is more to engineering than just "engineering" and be inspired to use their skills and mindset to practice social entrepreneurship and pursue ideas that make a difference.

Course Introduction

We are facing growing social and environmental challenges in our world where the solutions are not profitable financially but create significant social benefit and change. We must therefore create an environment where engineers have not only a social awareness, but also the skills and knowledge to build, work in and/or direct organizations where maximizing profits will also include maximizing worker, society and environmental benefits.

There are many examples of organizations involved in engineering, social change and philanthropy. Engineering for Change (E4C) and Engineers Without Borders (EWB) are good examples of nonprofit organizations directly involved in allowing engineers to use their skills in philanthropic ways. There are many major corporations, including Palantir, General Electric, Alcoa, Google, and many others, who are quite active in philanthropic work and have significant budgets to support these activities. Organizations such as the Gates Foundation utilize engineers with wide-ranging skillsets to take on significant global issues with aggressive goals, as do the U.N., RAND Corporation, the World Bank and others. There are an estimated 2.3 million operating nonprofits in the US, employing approximately 13.7M workers, which represents 10% of the nation's workforce. Consequently, engineers have a significant role to play in this environment.

Many engineering students are already interested in and excited by the possibilities of putting their developing engineering skills to use in ways that benefit humanity without financial gain, and that interest will continue throughout their careers. In recognition of this, we are offering this course in which students will:

- Understand the interaction among engineering, social change and philanthropy, and how organizations engage in these activities.
- Articulate their view and philosophy of engineering as it creates social change.
- Practice leadership, teamwork, entrepreneurial skills and decision making by awarding a significant financial contribution to a nonprofit foundation.
- Practice the art of multi optimization in an environment with severe cost restraints to support underfunded projects of significant social value.

An integral part of this course is student participation. Students will be required to take part in impromptu group discussions, team exercises and decision-making not only in the classroom, but also outside. Students will engage with on-campus and external organizations and their representatives as part of the process of evaluation and grant-making, and at the end of the semester, the students will make formal presentations in support of an idea for social change they believe can be successful.

Course Components

The course is comprised of the following main components:

1. Engineering and Social Change Lectures

Students will hear lectures from associated faculty on issues of current interest in the area of Engineering and Social Change. Topics covered are at the intersection of engineering, technology and society and include autonomy, sustainability, energy efficiency, low cost engineering, design using crowdsourcing and the future of engineering among others. We will also include a series of guest speakers of significant reputation in their fields, such as those hosted from the Gates Foundation and USAID in previous semesters, and these speakers will vary each semester.

2. Philanthropy Lectures

Students will hear lectures from course partners that will define philanthropy as an exploration of how one develops a vision of the public good and then deploys resources (including donations, volunteers, and engineering/invention) to achieve an impact.

3. $10,000 Neilom Foundation Engineering for Social Change Grant

Students will go through the rewarding process of granting $10,000 to a local nonprofit organization of their choice to support a "theme" of their choice. An initial list of themes will be presented and the class will vote for one theme they wish to support and identify a number of nonprofit organizations operating within the theme of interest. Substantial information will be collected on these non-profit organizations (history, mission statement, measures of success, budget, infrastructure, etc.) and the class will learn basic budgeting information in order to assess proposals. The class will then prepare a mission statement and a request for proposals (RFP) that will go out to nonprofits identified as working within the theme area and consistent with the class goals as noted in the RFP. A number of nonprofits will respond with proposals, and the students will review and rate all submitted proposals. The students will then participate actively in both phone interviews and site visits partially outside of class hours to selected nonprofit organizations. At the end of the course the class will decide, through technical analysis, discussion and voting, to grant $10,000 to one nonprofit due to a generous contribution from the Neilom Foundation, and on December 15th a final grant award ceremony will be held.

4. The Ideas for Social Change Challenge (ISCC)

The ISCC offers the students an opportunity to undertake an entrepreneurial approach to address a well-defined problem either in our local community or globally, utilizing knowledge they have gained about engineering, philanthropy and social change throughout the course. The students will work closely together in groups throughout the semester on their projects to give shape to their venture, and finally present their ISCC project at the end of the course, in addition to submitting a written project report.

5. Blog

Students are required to write two public blog posts on the class blog (see Canvas for details) during the semester describing their initial point of view surrounding the ideas of engineering, social change and philanthropy, and one blog post in the last week of the semester describing their experience and how this course has changed or impacted their viewpoint on engineering and social change.

6. Grant Ceremony

At the end of the semester students will be required to attend the grant ceremony, held on the same day as the scheduled day for a final exam for the class. During the ceremony the main grant award will be made to the successful nonprofit as chosen by the students, final ISCC presentations will also take place, and there will be a final celebration of the activities of the course.

Course Topics include:

- Philanthropy for Engineers
- Motivations for Doing Good
- From Idea to Action
- Autonomy
- Social Impact of Cryptocurrency
- Innovation and Entrepreneurship
- Low Cost Engineering and Codesign
- How Science and Technology Influences Human Behavior, and Vice-Versa
- Design Using Crowdsourcing
- The Future of Engineering
- Energy and Policy from an International Standpoint

Learning Outcomes (ABET)

- An understanding of professional and ethical responsibility
- The broad education necessary to understand the impact of engineering solutions in a global, economic, environmental, and societal context
- Knowledge of contemporary issues
- Ability to design a process to meet desired needs within realistic constraints such as economic, environmental, social, political ethical, health and safety and sustainability
- An ability to function on multidisciplinary teams

- An ability to identify, formulate, and solve engineering problems
- An ability to communicate effectively
- A recognition of the need for, and an ability to engage in, life-long learning
- Ability to use the techniques, skills, and modern engineering tools necessary for engineering practice.

Number of credits

3

Prerequisites

Academic Standing: Students with sufficient credit for senior standing or permission of the Mechanical Engineering Department.

Course duration

Fall 2015, September 1st to December 15th (date of final exam)

Class hours

Tuesdays and Thursdays 9:30 - 10:45 am, KEB Room 1200.

Technology Requirements

All students must come to class with an internet-connected device (phone, tablet or laptop), on which to perform voting tasks on during our philanthropy sections each week. We will be using Socrative.com for our anonymous class voting process – if you will use a mobile device for voting please make sure you have the free Socrative Student app available in the Apple or Google Play app stores.

Course Instructors

Professor Davinder K. Anand: Course Leader
Center for Engineering Concepts Development (CECD), Department of Mechanical Engineering, A. James Clark School of Engineering

Mr. Dylan Hazelwood: Course Manager
Center for Engineering Concepts Development (CECD), Department of Mechanical Engineering, A. James Clark School of Engineering

Dr. Mukes Kapilashrami: Course Instructor
Center for Engineering Concepts Development (CECD), Department of Mechanical Engineering, A. James Clark School of Engineering

Dr. Jennifer Littlefield: Course Instructor
Center for Philanthropy and Non-profit Leadership, School of Public Policy

Professor Robert Grimm: Course Instructor
Center for Philanthropy and Non-profit Leadership, School of Public Policy

Office Hours

Official office hours are in the Engineering for Social Change Lab, Room 2142, Glenn L. Martin Hall, Tuesday/Thursday at 2:00 - 4:00 pm. Other hours are by appointment; contact the Course Manager at dylan@umd.edu.

Textbook

The readings for the course are available through Canvas, or are freely available on the Internet. Case studies will be paid for and provided free of charge for students by the Center for Engineering Concepts Development (CECD) in the Department of Mechanical Engineering.

Course Website

Course announcements and all relevant information will be sent through Canvas, UMD's learning management system.

Course Elements and Grading

Course Element	Score	Assignment Type
Ideas for Social Change Challenge	30%	Group
Case Studies	10%	Individual
Proposal Review	10%	Individual
Midterm	15%	Individual
Engagement and Attendance	20%	Individual/Group
Statement of Interest	10%	Individual
Blog Post	5%	Individual

Total **100%**

Engineering for Social Change Lab

The Engineering for Social Change lab is a meeting room and office in Room 2142 in Glenn L. Martin Hall where students are encouraged to stop by for informal/formal discussions on all aspects of the course throughout the semester. There will also be periods for the ISCC project where students will be required to visit for feedback on their project – please see the ISCC project documentation for details on Canvas.

Exams

There will be a midterm exam representing 15% of the total grade, and the final exam at the end of the semester is comprised of two parts: (i) a written report of the semester-long ISCC project, and (ii) a group presentation of the project during the grant ceremony at the end of the semester during the final exam period.

Participation Grade

Student participation is required as this is an interactive course with student choice-driven milestones and activities. We will record an engagement and attendance grade during the semester comprising 20% of the final grade. Participation is comprised of classroom engagement and active discussion, as well as phone interviews and site visits with nonprofit organizations.

Course Etiquette

Attendance to both Tuesday and Thursday classes is **required**. In lectures and discussions we expect students to listen and respond to their peers respectfully. We request that students not use electronic devices unless relevant to the coursework or requested during the class period.

Note: *We strongly encourage the students to come prepared to class by looking into the specific topical area for the week in order to actively participate in class discussions. It is recommended that each student read the biographies and visit the websites of the speakers in order to achieve a better understanding for their respective field of expertise.*

Course Timeline

Week	Tuesday	Topic	Thursday	Topic
1	1-Sep	Prof. Anand *Introduction to ESC*	3-Sep	Prof. Grimm *Philanthropy for Engineers*
2	8-Sep	Dr. Kapilashrami *Introduction to ISCC*	10-Sep	Prof. Grimm *Motivations for Doing Good*
3	15-Sep	*From Idea to Action G. Pastor, J. Claggett, Mukes Kapilashrami*	17-Sep	Dr. Ryan Shelby *USAID*
4	22-Sep	Dr. Littlefield and Mr. Hazelwood - *Mechanics of Voting*	24-Sep	Dr. Littlefield and Mr. Hazelwood – *Initial Theme selection*
5	29-Sep	Erica Estrada-Liou *Innovation and Entrepreneurship*	1-Oct	Dr. Kapilashrami *ISCC group presentation: Executive Summary*
6	6-Oct	Prof. Fuge *Design via crowdsourcing*	8-Oct	Dr. Littlefield and Mr. Hazelwood *Mission, Budgets*
7	13-Oct	Prof. Pecht *Does Science and Technology influence behavior*	15-Oct	Dr. Littlefield and Mr. Hazelwood *Grant Process: RFP Part 2*
8	20-Oct	Dr. Firebaugh *Autonomy*	22-Oct	Prof. Vaughn-Cooke *Human Factors Engineering*
9	27-Oct	Dr. Kapilashrami *ISCC Workshop*	29-Oct	Andrew Miller *Social Impact of Encrypted Currency*
10	3-Nov	Dr. Littlefield and Mr. Hazelwood *Voting for Phone Interviews*	5-Nov	Dr. Littlefield, Mr. Hazelwood *Grant Process: Phone Interviews*

Course Timeline (Continued)

Week	Tuesday	Topic	Thursday	Topic
11	10-Nov	Smeeta Hirani *From Microsoft to Social Change*	12-Nov	Dr. Littlefield and Mr. Hazelwood *Voting - Site Visits*
12	17-Nov	Dr. Kapilashrami *ISCC final draft presentations*	19-Nov	Mr. Hazelwood *Site Visits*
13	24-Nov	Prof. Forman *Water and Sustainability*	26-Nov	Thanksgiving
14	1-Dec	Prof. Kim *Future of Engineering*	3-Dec	Prof. Gabriel *International Energy and Policy*
15	8-Dec	Dr. Littlefield and Mr. Hazelwood *Final Vote*	10-Dec	Prof. Anand What have we learned?

Appendix H

Entrepreneurship in Southern Maryland Challenge

As part of an ongoing partnership with the College of Southern Maryland (CSM) and an expansion of CECD's Engineering for Social Change efforts, CECD supported 25 students in BAD 1210, a business administration course with an innovative co-developed Entrepreneurship in Southern Maryland Challenge. The challenge required students to work in teams to undertake an entrepreneurial approach to identify a problem pertinent to their community in the Southern Maryland region. After feedback from CECD staff and direction from course leader Dr. Mary Beth Klinger of CSM, the students then developed a sustainable solution using their management skills, with an emphasis on creating maximum social impact. A celebration was held on May 5, 2016 at the La Plata, Maryland campus, with Senator Thomas "Mac" Middleton in attendance, as well as CSM President Bradley Gottfried, faculty, media and nonprofit representatives, students and family members. The Neilom Foundation provided $2,000 in total prizes. The $1,500 First Place prize went to the Team Life Planning Curriculum Project, with the $500 Second Place prize going to Team Planting Hope, who decided to donate their winnings to the nonprofit they worked with, the Southern Maryland Food Bank. A brief summary of student projects is below.

Group One: Life Planning Curriculum Project

This student team created a Life Planning Curriculum for implementation in the local public school system in St. Mary's County, Maryland. The project integrated four critical components the group believed was essential to successful post High School graduation outcomes. The group proposed an evidence-based method for converting student potential into real world success stories within St. Mary's County. The students planned the implementation of an integrated curriculum starting in the 8th grade and going through the 12th grade. The premise was to help youth develop essential non-cognitive skills along with the practical life skills needed for success post High School graduation. Their project idea strove to help bridge the current gap between cognition, affect (the experience of emotion), and human physiology (the body's response) in an effort to help these young people ultimately be more successful personally, academically, and professionally. At the end of the semester the group had planned to meet with multiple school administrators to begin the process of adopting their curriculum in after school programs.

Group Two: The Mission Project

Students in this group chose to positively impact the homeless community through a series of actions in conjunction with The Mission, a local organization providing shelter to the homeless in St Mary's County, Maryland. The team encouraged community members to support the homeless in their community, setting an example with their own time and funds. They supported the mission by attending meetings, passing out flyers, volunteering at a day center and providing free audio/visual services to the organization. The student team also supported The Mission by donating the group's $250 seed funding. Team members organized meetings with local businesses and churches, successfully involving two local businesses and convincing them to attend volunteer meetings, as well as encouraging a church to send over fifty men to attend a volunteer meeting. The students created a student service learning opportunity at CSM and successfully signed up their chosen organization as a trusted organization for further student engagement in future.

Group Three: Food Bank Garden Project

This student team worked with the Southern Maryland Food Bank to develop a community garden project to grow fruits and vegetables to supply to those in need. The concept was that following an initial investment, community efforts would ensure the garden would thrive and continue to provide essentially 'free' food to the hungry. More than just being a garden for supplies the garden would include flowers, honeybees, and a teaching space, which would be open to any homeschooled students in the Southern Maryland tri-county area. The student team undertook the simultaneous challenges of developing the garden with their project seed funds and trying to raise awareness and further financial support. The students targeted locally owned companies for donations, approached law offices and insurance companies for sponsorships, and greeneries and nurseries for supplies. By the conclusion of the semester, the garden space construction was well underway, with the students having secured five sponsorships for the garden, as well as sponsors for three children through the organization's Snack Sak program.

Group Four: Higher Heights Foundation

This student group focused on the issue of adolescents who are post-high school but unable to go to college, with the plan to develop a non-profit foundation to provide assistance. The group hoped to engage the help of the community and CSM, providing these students with paid tuition through multiple sources such as CSM, county government, and local businesses. It was proposed that in return these students would provide their time and efforts to address local community issues. To be sure the opportunity of a college education was not wasted, the group designed a set of requirements for program participants. Students would have to maintain a 2.5 GPA, receive no criminal charges, attend foundation meetings, and get letters of reference from their professors. The hope was that students who were successfully assisted through this foundation would join the effort, and help to assist the next generation of students.

Group Five: Save The Bay

Maryland's Chesapeake Bay is a popular and important natural resource for those who live in the region, providing food, water, a transportation route, recreation, and scenic views. The bay also deals with issues of pollution and a resulting decline in sea life. The pollution causes algae to bloom, blocking light and removing oxygen from the water, making it harder for the wildlife to live. As a result of overfishing of the oyster population in the Bay, the lack of these effective water-filtering creatures has allowed pollution to inflict serious damage. Oysters can effectively filter out pollutants such as nitrogen and phosphorus from up to 50 gallons of water a day, keeping algae blooms in check. The challenge is how to get the oysters back up to normal levels. The student team suggested the institution of a five-year buyout of oyster farmers, similar to the successful tobacco farmer buyout in the region decades ago. The students suggested the current ten million dollar excess in the St. Mary's County budget could facilitate their plan, allowing the population to return to the numbers needed to effectively filter the water of the Bay. Alternatives were also suggested in their plan, such as oyster farming. Alternatives such as these are predicted to allow harvesting at current levels while keeping the natural population high.

Appendix I

A Survey of Other Institutions

A number of forward-looking institutions have begun to integrate engineering and social impact thinking into their curriculum in various forms. While by no means an exhaustive list, we have identified some key programs at other universities working at the intersection of engineering and social change.

1. University of California, Berkeley: Designated Emphasis in Development Engineering Program [435]

The Designated Emphasis in Development Engineering is an interdisciplinary program offered by the Department of Civil and Environmental Engineering at UC Berkeley. It is specifically meant for Doctoral students whose research encompasses the use of technology and engineering to tackle the needs of the poverty stricken people all over the globe. The program equips students with the skill to take initiative, devise and intervene in technological development in complex and less resourceful situations.

The course structure of the program includes 2 core subjects, 3 electives, and a dissertation in alignment to the program goals. There are about 22 faculty members catering the students. Additionally, with the help and coordination of affiliates like the Blum Center of Development Economics, BEST lab, Development Impact lab, etc., students get access to a variety of resources in their intellectual pursuits. Ongoing research in the program includes human-centric design, impact analysis, remote sensing, data science and more.

2. University of California, Berkeley: BEST lab [436]

The BEST lab at University of California, Berkeley, directed by Mechanical Engineering professor Alice Agogino, conducts research in computational design, gender equity, human-machine cognition, supervisory control, co-robotics, sensor fusion, design research and intelligent learning systems. Current projects include the Berkeley Prosthetics Project (BPP), where a team of volunteers designs custom prosthetics for those in need.

3. Michigan Technology University: D80 program [437]

The D80 program was established in 1996, and is driven by the mission to develop solutions to the problems faced by the vulnerable people who constitute about 80% of the earth's human population. The programs offered by D80 include certifications such as the International Sustainable Development Engineering (ISDE) Certificate, which can be acquired simultaneously while pursuing any engineering baccalaureate program. Additionally, it also encompasses research groups like Michigan Tech Open Sustainability Technology (MOST), which conducts research on projects such as solar energy to provide a clean sustainable source of electricity.

4. University of Minnesota: Acara Program [438]

The Acara program is run by the Institute of Environment, College of Engineering and Sciences, and Carlson School of Management at the University of Minnesota. Its aim is to develop scalable entrepreneurial solutions to curb global community problems. It believes in the combined effort of education and impact. Some of the courses it offers include Global Venture Design, Design for Sustainable Development, etc. Since its inception the Acara program has tutored more than 1,000 entrepreneurs who are pursuing more than 100 different ventures to positively impact society.

5. Stanford University: D School [439]

The D School is described as a congregation of engineers from various fields, brought together to solve problems with humanitarian values at the core of their operation. It does not provide degrees or grants, but the multidisciplinary courses offered by the D School can be integrated with the Stanford Design Program or any graduate degree. They strongly believe in collaborative efforts and partner with corporate, non-profit and government organizations to develop real-world projects. According to the D School, "We believe that everyone is a student of innovation, with a responsibility to nurture the potential for innovation in others."

6. Stanford University - EXTREME lab [440]

EXTREME is a multidisciplinary course offered as a collaborative effort by the Graduate School of Business and the School of Mechanical Engineering at Stanford University. The goal of the course is to cultivate an ecosystem, which will enable students to develop products and services directed towards the needs of the poorest citizens of the world. Since the inception of the program nine years ago, 80 highly effective projects have been deployed in 14 different countries. Students collaborate with five global partners each term to create innovative and low cost solutions. They are assisted by the course's teaching team, alumni, and other experts in the process of design and accessing resources. Once the project is approved, the solution is implemented by the partners or relevant organizations. Example projects include the development of a uniform seed-planting tool for use in Myanmar to improve plant yield, and improvements in efficiency and cost for water filter production in Guatemala.

7. MIT – D Lab [441]

MIT's D Lab stands for "development through discovery, design and distribution". The lab's programs include a variety of courses that enable students to innovate and implement solutions that benefit people living under the influence of poverty. The lab deals with topics like health, energy, agriculture, waste management, supply chain management, business development etc., focusing on initiating real-time projects with community development and scalability. Some of the current projects at the D Lab include water purification, refrigeration, and low cost methods of environmental sensing.

8. MIT - GEAR Lab [442]

The Global Engineering and Research (GEAR) Lab, directed by Prof. Amos Winter, is a part of the Mechanical Engineering Department at MIT. It incorporates mechanical concepts and user-centric design philosophy to develop technologies that constructively influence the society. One of the Lab's popular projects is the Leverage Freedom Chair (LFC). It is an all-terrain wheelchair, developed to function reliably on unpaved roads in rural areas with simple, easy to replace, commonly available components.

9. Carnegie Mellon University: Engineering & Public Policy Department [443]

The Engineering and Public Policy Department (EPP) at Carnegie Mellon University is motivated to solve technology-based policy issues. Through its multidisciplinary coursework, EPP imparts strong engineering skills and a wider perspective to students, making them sound professionals. It conducts research on problems that are experienced in technical innovations, risk analysis, energy, environmental systems, information technology and policymaking. The EPP has 53 faculty and 7 research staff members and offers Ph.D., double major undergraduate, and Master's degrees. As an extension to EPP's on-campus location, the DC office (located in Washington, DC) coordinates activities involving participation in the policymaking processes of the government. This helps the students get much needed practical exposure.

10. MIT: Technology and Public Policy Program [444]

The Technology and Public Policy (TPP) Program at MIT strategizes the process of policy making to be more intelligent and informed by incorporating concepts from engineering and science. The Multidisciplinary coursework of TPP includes topics in law, economics, engineering, policy, and a thesis on technological policy issue.

11. Stanford University: Management Science and Engineering Department [445]

The Management Science and Engineering Department at Stanford University was established in 1999 with the aim of interfacing engineering, business, and public policy. Since its inception, the Department has developed to support a variety of topics for research like Information Technology, Probabilistic and Stochastic Models, Economics and Finance, Risk Analysis, etc.

12. Purdue University: Global Engineering Programs [446]

The Global Engineering Programs at Purdue University strive to increase the involvement of engineering students globally, bolster the global sustainability movement, and encourage international partnerships for collaborative research.

References

[432] Alan P. Santos/DC Sports Box, "ENME 467 Grant Celebration", Center for Engineering Concepts Development, December 17, 2015.
[433] Val Nyce, "BAD 1210 Celebration", College of Southern Maryland, May 5, 2016.
[434] Lisa Helfert, "ENME 467 Spring Celebration", Center for Engineering Concepts Development, May 18, 2015.
[435] "Development Engineering", University of California, Berkeley, 2016, accessed at http://deveng.berkeley.edu
[436] "BEST Lab UC Berkeley", University of California, Berkeley, 2016, accessed at http://best.berkeley.edu
[437] "D80 prosperity by design", Michigan Tech, 2016, accessed at http://www.mtu.edu/d80
[438] "Acara Impact Entrepreneurship", University of Minnesota, 2016, accessed at http://acara.environment.umn.edu
[439] "D.School", Stanford University, 2016, accessed at http://dschool.stanford.edu/our-point-of-view
[440] "Extreme", Stanford University, 2016, accessed at http://extreme.stanford.edu
[441] "D-lab", Massachusetts Institute of Technology, 2016, accessed at http://d-lab.mit.edu
[442] "Gear Program", Massachusetts Institute of Technology, 2016, accessed at http://gear.mit.edu
[443] "Engineering and Public Policy", Carnegie Mellon University, 2016, accessed at https://www.cmu.edu/epp
[444] "MIT Technology and Policy Program", Massachusetts Institute of Technology, 2016, accessed at http://tppserver.mit.edu
[445] "Management Science & Engineering", Stanford University, 2016, accessed at http://msande.stanford.edu/about/department-overview
[446] "Global Engineering Programs", Purdue University, 2016, accessed at https://engineering.purdue.edu/GEP/About/mission